The World Bank Policy for Projects on International Waterways

An Historical and Legal Analysis

I0042410

Law, Justice, and Development

The Law, Justice, and Development series is offered by the Legal Vice Presidency of the World Bank to provide insights into aspects of law and justice that are relevant to the development process. Works in the series present new legal and judicial reform activities related to the World Bank's work, as well as analyses of domestic and international law. The series is intended to be accessible to a broad audience as well as to legal practitioners.

Series Editor: Salman M. A. Salman
Editorial Board: Hassane Cisse, Alberto Ninio, and Kishor Uprety

The World Bank Policy for Projects on International Waterways

An Historical and Legal Analysis

Salman M. A. Salman
Lead Counsel
Legal Vice Presidency
The World Bank

THE WORLD BANK
Washington, DC

ISBN: 978-0-8213-7953-0
eISBN: 978-0-8213-7954-7
DOI: 10.1596/978-0-8213-7953-0

Library of Congress Cataloging in Publication Data

Salman, Salman M. A., 1948-
The World Bank policy for projects on international waterways : an historical and legal analysis / Salman M.A. Salman.
     p. cm. — (Law, justice, and development)
     Includes bibliographical references and index.
     ISBN 978-0-8213-7953-0 — ISBN 978-0-8213-7954-7
     1. International rivers. 2. Waterways—International cooperation. 3. Water rights (International law) 4. Economic development projects—Law and legislation. 5. Water resources development—Law and legislation. 6. Water resources development—International cooperation. 7. World Bank. I. World Bank. II. Title.

HE385.S25 2009
387—dc22

2009011998

Contents

APPENDICES

Foreword

More than 300 rivers and 100 lakes are shared by two or more states. These basins cross the territories of 145 states, and cover about half of the earth's surface. About 40 percent of the world's population lives around these basins. Together with the close to 300 transboundary aquifers, these watercourses bestow humanity with a precious resource that has no substitute, and upon which there is a total dependence for survival.

Notwithstanding these pressing facts, shared watercourses remain one of the most important resources without a global treaty in force regulating their sharing, use and protection. The United Nations Convention on the Law of the Non-Navigational Uses of International Watercourses was adopted by the United Nations General Assembly in 1997, after about 27 years of preparatory work. However, 12 years after its adoption, the Convention has yet to command sufficient ratifications to enter into force and effect. Moreover, not many watercourses are covered by agreements between the states sharing them, and even when such agreements exist, they, in most cases, do not encompass all the riparians. As a result, a number of the shared watercourses are characterized by existing or emerging disputes involving some or all of the riparian states. Climate change and the increasing water consumption and demand due to the rapid population growth threaten to aggravate water scarcity, and in turn generate new disputes over shared watercourses, as well as exacerbate existing ones.

About 15 percent of the World Bank's annual lending program is for water supply, irrigation and hydropower projects, and many of those projects affect shared watercourses. Indeed, from the very early years of its operations, the Bank has faced the delicate and complex challenge of projects on international waterways, and the need to balance the interests of all the riparians. That challenge was compounded by the paucity of rules of international law in the field of shared watercourses. Confronted with this situation, the Bank had to reflect carefully on the matter, and develop approaches to these projects, keeping in mind the need to act in the interests of all its members. Consequently, in 1956, only 10 years after the Bank commenced its operational work, it issued the first policy for projects on international waterways. That policy has evolved through the years, and has been periodically updated and refined to reflect the Bank's extensive experience with projects in this field, and with the evolving principles of international water law.

Today the Bank policy stands out as the only comprehensive instrument relating to projects on international waterways with a global reach and application. The policy covers members of the Bank as well as non-members, and developing as well as developed countries. The policy, no doubt, is helping fill major gaps in international water law emanating from the absence of a global treaty in force on shared watercourses.

This Study traces the evolution of the Bank policy for projects on international waterways from the early years of Bank operations in the late 1940s until today. It provides a critical analysis of the legal context of the policy, and relates that context to the prevailing principles of international water law at the particular time of the policy issuance and updates. It illustrates how the policy has been adapted to the changing Bank lending program, and to the growing complexity of the projects that the Bank finances in this field.

The Legal Vice Presidency of the World Bank is pleased to offer this publication and hopes it will contribute to a better and greater understanding of the Bank policy for projects on international waterways, and the Bank's role and efforts in facilitating the establishment of cooperative arrangements for the efficient utilization and management of such waterways.

Scott B. White
Acting Senior Vice President and
General Counsel
The World Bank
January 2009

Abstract

This Study deals with the evolution and context of the Bank policy for projects on international waterways. It starts with a brief description of how the Bank has faced the challenges stemming from such projects, and the different approaches deliberated by the Bank that led to the issuance of the first policy in 1956. The Study then reviews the implementation experience of the policy and analyzes the principles and procedures, as well as the main features of each of the policies issued in 1956, 1965 and 1985, respectively. The principles of international water law prevailing at each stage of the policy issuance are examined and compared with those of the Bank policy. The Study also discusses in detail the notification process: its basis, by whom, to whom, its content, different riparians' responses, and the exceptions to the notification requirement. It then analyzes how the Bank handles an objection from one or more of the riparians to projects proposed for Bank financing. It also examines how the Bank has dealt with transboundary groundwater, as well as the linkages between the policy for projects on international waterways and the policies on disputed areas and environmental impact assessment. The conclusion provides an overview of the main findings of the Study, and highlights some of the lessons drawn from the implementation experience of the policy.

Acknowledgments

A number of colleagues have assisted in one way or another in the preparation of this Study. Some have helped in the research of Bank files, or of treaties and other international instruments, while others have read and provided valuable comments on one or more chapters of the Study, or some of the principles discussed therein, and I am grateful to all of them. In particular, I would like to extend my sincere thanks and appreciation to Götz Reichert, Suryna Ali, Fuad Bateh, Shéhan de Sayrah, Sachiko Morita, Yuan Tao, Kerstin Mechlem, Gabriel Eckstein, Ousmane Dione, Inger Andersen, Malcolm Cosgrove-Davies, Stephen Lintner, Vahid Alavian, Robert Schlotterer, Raghuveer Sharma, John Kittridge, Ashok Subramanian, Joop Stoutjesdijk and Jaap van Opstal.

I am also grateful to David Freestone, Charles Di Leva, Hadi Abushakra, Hassane Cisse, Alberto Ninio, Kishor Uprety, Evarist Baimu and David Grey for their helpful comments on earlier versions of the manuscript; and to Scott White for his enthusiastic support and advice throughout the preparation of the Study.

My thanks are also due to Laura Lalime-Mowry, Christian Jimenez, Olesya Zaremba, Maria Luisa Koransky, Jose Santos, Richard Carter, Manuel Rodrigues, Lakshmi Mathew, Mary May Agcaoili and Sarah Salman for assistance in various ways with the Study.

Acronyms and Abbreviations

ADB	Asian Development Bank
AfDB	African Development Bank
AMS	Administrative Manual Statement
APL	Adaptable Program Loans
BNWPP	Bank-Netherlands Water Partnership Program
BP	Bank Procedures
CBD	Convention on Biological Diversity
CDD	Community Driven Development
CEP	Caspian Environment Program
DPL	Development Policy Lending
EAC	East African Community
EBRD	European Bank for Reconstruction and Development
EC	European Commission
EEZ	Exclusive Economic Zone
EIA	Environmental Impact Assessment
ERP	European Recovery Program
FAO	Food and Agriculture Organization of the United Nations
GEF	Global Environment Facility
GP	Good Practices
IBRD	International Bank for Reconstruction and Development
ICJ	International Court of Justice
ICPDR	International Commission for the Protection of the Danube River
ICPOAP	International Commission for the Protection of the Odra against Pollution
ICSID	International Centre for Settlement of Investment Disputes

ICWC	Inter-state Commission for Water Coordination
IDA	International Development Association
IDB	Inter-American Development Bank
IFAS	International Fund for Saving the Aral Sea
IFC	International Finance Corporation
IHP	International Hydrological Programme
IIL	Institute of International Law (*Institut de Droit International*)
ILA	International Law Association
ILC	International Law Commission
IWRM	Integrated Water Resources Management
LCBC	Lake Chad Basin Commission
LVBC	Lake Victoria Basin Commission
MEAs	Multilateral Environmental Agreements
MICs	Middle-Income Countries
MIGA	Multilateral Investment Guarantee Agency
MRC	Mekong River Commission
NBA	Niger Basin Authority
NBI	Nile Basin Initiative
NGOs	Non-Governmental Organizations
NWSAS	North Western Sahara Aquifer System
OD	Operational Directive
OM	Operational Manual
OMS	Operational Manual Statement
OMVS	*Organisation pour la mise en valeur du fleuve Sénégal*
OP	Operational Policies
OSS	*Observatoire du Sahara et Sahel*
PAD	Project Appraisal Document
PCA	Permanent Court of Arbitration
PCIJ	Permanent Court of International Justice
PID	Project Information Document
POPs	Persistent Organic Pollutants
SADC	Southern African Development Community
SAR	Staff Appraisal Report

SDN	Sustainable Development Network
SEA	Social and Environmental Assessment
SSEA	Strategic/Sectoral Social and Environmental Assessment
UNCCD	United Nations Convention to Combat Desertification
UNCLOS	United Nations Convention on the Law of the Sea
UNDP	United Nations Development Programme
UNECE	United Nations Economic Commission for Europe
UNEP	United Nations Environment Programme
UNESCO	United Nations Educational, Scientific and Cultural Organization
UNFCCC	United Nations Framework Convention on Climate Change
UNGA	United Nations General Assembly
UNMIK	United Nations Mission in Kosovo
WBG	World Bank Group
WCD	World Commission on Dams
WCW	World Commission on Water for the 21st Century
WWC	World Water Council

Cases

Case Concerning Kasikili/Sedudu Island (Botswana/Namibia) ICJ, 1999.
Case Concerning Territorial and Maritime Dispute between Nicaragua and Honduras in the Caribbean Sea (Nicaragua v. Honduras) ICJ, 2007.
Case Concerning the Gabčíkovo-Nagymaros Project (Hungary v. Slovakia) ICJ, 1997.
Case Relating to the Territorial Jurisdiction of the International Commission of the River Oder (United Kingdom, Czechoslovakia, Denmark, France, Germany and Sweden v. Poland) PCIJ, 1929.
Corfu Channel case (U.K. v. Albania) ICJ, 1949.
Dispute Regarding Navigational and Related Rights (Costa Rica v. Nicaragua) ICJ [Pending Decision].
Diversion of Water from the Meuse (Netherlands v. Belgium) PCIJ, 1925.
Eritrea-Ethiopia Boundary Commission, PCA, 2002.
Frontier Dispute (Benin/Niger) ICJ, 2005.
Lake Lanoux Arbitration (France v. Spain) 24 I.L.R., 1957.
Land and Maritime Boundary between Cameroon and Nigeria (Cameroon v. Nigeria: Equatorial Guinea intervening) ICJ, 2002.
Maritime Delimitation in the Black Sea (Romania v. Ukraine) ICJ, 2009.
Oscar Chinn case (Great Britain v. Belgium) PCIJ, 1934.
Pulp Mills on the River Uruguay (Argentina v. Uruguay) ICJ [Pending Decision].
Temple of Preah Vihear (Cambodia v. Thailand) ICJ, 1962.
Trail Smelter Arbitration (United States v. Canada) R.I.A.A., 1941.

CHAPTER 1

Facing the Challenges of International Waterways

1.1 The Bank's Early Years

The Articles of Agreement of the International Bank for Reconstruction and Development (IBRD, or the Bank)[1] were discussed and agreed upon during the meeting that was held in Bretton Woods, New Hampshire, in July 1944, and they entered into force on December 27, 1945.[2] The first meeting of the Executive Directors of the Bank took place on May 7, 1946, following the meeting of the Board of Governors of the Bank in March 1946.[3] The Bank opened for business on June 25, 1946, two years after the Bretton Woods meeting, with its membership reaching 38 states.[4]

---

[1] A distinction needs to be made at the outset between the terms "IBRD", "The World Bank" and "The World Bank Group." The term "The World Bank" refers to both, the "International Bank for Reconstruction and Development (IBRD)" and the "International Development Association (IDA)." As will be discussed later, IDA was established in 1960 to provide interest-free credits to its poorest member countries. IBRD operations are concentrated in middle-income countries (MICs) and creditworthy poor countries. The wider term "The World Bank Group (WBG)" includes, in addition to the Bank and IDA, the International Finance Corporation (IFC), the Multilateral Investment Guarantee Agency (MIGA), and the International Centre for Settlement of Investment Disputes (ICSID). For more details on each of those institutions, *see infra*, Chapter 3, Part 3.3 of this Book. *See* also *The World Bank Annual Report* 2008, available at: http://web.worldbank.org /WBSITE/ EXTERNAL/EXTABOUTUS/EXTANNREP/EXTANNREP2K8/0,,content MDK:21918085~menuPK:5405445~pagePK:64168445~piPK:64168309~theSitePK:516 4354,00.html. For the Articles of Agreement of each of these institutions, *see*: http:// web.worldbank.org/WBSITE/EXTERNAL/EXTABOUTUS/ORGANIZATION/BODEXT/ 0,,contentMDK:50004943~menuPK:64020045~pagePK:64020054~piPK:64020408~the SitePK:278036,00.html.

[2] For a detailed account of the history of the negotiations of the Articles of Agreement of the Bank, *see* Edward S. Mason & Robert E. Asher, *The World Bank since Bretton Woods*, Chapter 1 (The Brookings Institution 1973).

[3] The Articles of Agreement vest all powers of the Bank in the Board of Governors which consists of one governor for each member country. The voting power of each member is determined by the shares it holds in the Bank. This weighted voting system distinguished the Bank, from the start, from other international organizations where each member has one vote. The Board of Governors selects the Executive Directors and has delegated to them a general authority to carry out the day-to-day operations of the Bank.

[4] *See First Annual Report* by the Executive Directors, Submitted to the First Annual Meeting of the Board of Governors of the International Bank for Reconstruction and Development,

A year later, on August 15, 1947, the Bank concluded an agreement with the United Nations whereby the Bank's independence of the United Nations system was established.[5] The Agreement stated that "By reason of the nature of its international responsibilities and the terms of its Articles of Agreement, the Bank is, and is required to function as, an independent international organization."[6] Consequently, the Agreement recognized that the action to be taken by the Bank on any loan is a matter to be determined by the independent exercise of the Bank's own judgment in accordance with the Bank's Articles of Agreement.[7] Thus, the Bank became an autonomous "specialized agency" of the United Nations, with no reporting obligations to, or oversight authority from the United Nations.[8]

One main purpose of the Bank, as set forth in its Articles of Agreement, is to "assist in the reconstruction and development of territories of members by facilitating the investment of capital for productive purposes, including the restoration of economies destroyed or disrupted by war, the reconversion of productive facilities to peacetime needs and the encouragement of the development of productive facilities and resources in less developed countries."[9] The emphasis, as such, was on reconstruction and restoration of the war-affected economies.

The Bank spent most of its energy and time in its early years trying to establish some basic rules, procedures, and a structure for conducting its business, and to clarify the relationship between the President of the Bank and its Executive Directors. The development of procedures for loan applications was one major

---

September 27, 1946, at 1. For the membership of the Bank, *see id.*, at 1. *See* also Mason & Asher, *supra* n. 2, at 35–36.

[5] *See* Agreement between the United Nations and the International Bank for Reconstruction and Development, 16 U.N.T.S. 346 (1948); also available at: http://treaties. un.org/untc//Pages//doc/Publication/UNTS/Volume%2016/volume-16-II-109-English.pdf. The Agreement entered into force on November 15, 1947, following its approval by the Bank's Board of Governors on September 15, 1947, and by the United Nations General Assembly on November 15, 1947. *See* Protocol attached to the Agreement.

[6] *See id.,* Article (I)2.

[7] *See id.,* Article IV(3). This Article goes on to state that "the United Nations recognizes, therefore that it would be sound policy to refrain from making recommendations to the Bank with respect to particular loans or with respect to terms or conditions of financing by the Bank."

[8] It should be clarified, however, that Article VI of the Agreement states that "The Bank takes note of the obligation assumed, under paragraph 2 of Article 48 of the United Nations Charter, by such of its members as are also members of the United Nations, to carry out the decisions of the Security Council through their action in the appropriate specialized agencies of which they are members, and will, in the conduct of its activities, have due regard for decisions of the Security Council under Articles 41 and 42 of the United Nations Charter."

[9] *See* Article 1(i) of the Articles of Agreement of the Bank, *supra* n. 1.

challenge that the new Bank faced. Indeed, the studies conducted in the first year for such procedures were preliminary in nature, "since development of definitive policies in substantial detail can only be intelligently undertaken in the light of loan applications which shall be received."[10] The sources and manner of obtaining the funds required by the Bank for its operations, to supplement the paid-up capital, was another challenge that the Bank faced.

The relationship between the President of the Bank and the Executive Directors (or what is commonly known as the "Board of Directors," or simply "the Board") was another difficult issue that the Bank faced in its early years. The first President of the Bank, Mr. Eugene Meyer, "was confronted by a strong board of Executive Directors, led by the young, well informed, energetic and ambitious U.S. executive director. Much of Meyer's time was spent in battling with the board for leadership of the institution. His own previous experience in domestic financial agencies did not help him particularly in understanding balance-of-payments crisis, foreign exchange versus local currency requirements, and some of the other unprecedented problems that would have to be dealt with by the Bank."[11] Those problems culminated in the resignation of Mr. Meyer on December 4, 1946, about six months after he took office. His resignation was a clear indication of the difficulties that the Bank faced in its first few months. The Vice President, Mr. Harold Smith, took over as Acting President, but he died suddenly on January 23, 1947, thus serving less than two months. The Executive Directors elected Mr. John McCloy as President of the Bank, and he took office on March 17, 1947.[12] However, the struggle between the Executive Directors and the President for the leadership of the Bank continued, and the relationship between the two worsened further. As a result, Mr. McCloy was forced to resign two years later. Mr. Eugene Black was elected as President on May 18, 1949.[13] The consensus over his selection, and the manner in which he handled, from the very beginning, the relationship with the Executive Directors brought the needed stability, and helped the Bank to achieve "the kind of internal harmony that is essential to external effectiveness."[14]

The Bank approved its first project on May 9, 1947, almost a year after it formally opened for business. A loan of $250 million was granted to *Credit National*, a French corporation, with a guarantee from the French Government.

---

[10] *See supra* n. 4, *First Annual Report*, at 9.

[11] *See* Mason & Asher, *supra* n. 2, at 42.

[12] *See* International Bank for Reconstruction and Development, *Second Annual Report*, 1946–47, at 20–21.

[13] *See id.*, at 60–61.

[14] *See id.*, at 61.

The purpose of the loan was to assist in financing reconstruction and development of the French economy.[15] This loan was followed three months later by a loan of $195 million to the Kingdom of the Netherlands. The proceeds of that loan were to be used for the reconstruction of the productive facilities in the Netherlands.[16] Two more loans were granted in 1947 to the Kingdom of Denmark and the Grand Duchy of Luxembourg, respectively. The first loan was for assisting Denmark in its economic recovery program by financing the import of essential capital goods and raw materials. Similarly, the second loan was for the purchase of equipment for the Luxembourg steel industry and of rolling stocks for its railways.[17] These four loans, which were granted for assisting in the reconstruction of the recipient countries, were clearly made for financing imports—or what is known as "program lending"—as opposed to financing specific defined investment projects.[18]

The first developing country to receive a loan from the Bank was Chile. On March 25, 1948, the Bank extended two loans to Chile totaling $16 million. The first loan was to provide foreign exchange for the construction of additional hydroelectric plants and related transmission lines, installation of additional generating units and provision of pumping equipment for irrigation. The second loan was for the purchase of agricultural machinery.[19] Those loans were followed in 1949 by two loans to Mexico. The first loan was for purchase by the Federal Electricity Commission of Mexico of equipment and materials comprising new steam and hydroelectric generating stations, transmission lines and distribution systems for a number of its projects in various parts of Mexico.[20] The second loan to Mexico was also for electric power development, but was limited to Mexico City.[21] A loan was extended to Brazil in 1949 for expansion of its hydroelectric power and telephone facilities.[22] The loan was intended to

---

[15] *See supra* n. 12, at 18.

[16] *See id.*, at 18.

[17] *See* International Bank for Reconstruction and Development, *Third Annual Report, 1947–48*, at 21–22.

[18] As discussed earlier, the purposes of the Bank, as stipulated in Article 1(i) of its Articles of Agreement, emphasized the reconstruction of the territories devastated by the war. This Article was interpreted in 1946 to give the Bank authority to make or guarantee loans for "programs of economic reconstruction and the reconstruction of monetary systems, including long term stabilization loans." *See* Ibrahim Shihata, *The World Bank in a Changing World: Selected Essays and Lectures*, 25 (Martinus Nijhoff 1991).

[19] *See supra* n. 17, at 22–23.

[20] *See* International Bank for Reconstruction and Development, *Fourth Annual Report, 1948–49*, at 16–17.

[21] *See id.*, at 17.

[22] *See id.*, at 18.

assist in expansion of power-generation and distribution equipment, and the necessary additions to the facilities supplying water for power-generation purposes.[23] These loans, like the European loans, were also for general programs, and were not for specific projects. However, unlike the European loans, they were for development and not reconstruction purposes. It is worth noting that the Bank paid particular attention to the hydropower sector in its early lending operations to developing countries: four out of its first five loans were to assist in the generation of hydropower.

The Bank continued its programmatic lending for reconstruction in Europe by granting a loan to Belgium in 1949 for financing imports of equipment for the construction of steel mills and a power-generating plant.[24] Finland received a loan in 1949 for the purposes of completing part of its power program, as well as for modernizing its woodworking industry.[25] The emphasis on financing hydropower included both European and developing member countries.

The European Recovery Program (ERP, or the Marshall Plan) came into existence with the enactment by the United States of America of the Economic Cooperation Act on April 3, 1948.[26] Consequently, the European countries that benefited from the Marshall Plan were no longer totally dependent on the Bank for financial assistance. As a result, the Marshal Plan allowed the Bank to pay more attention to the development needs of its non-European member countries by freeing some of its resources for those countries.[27] As a result of the Marshall Plan, the Bank was able to turn from the emergency of reconstruction to its long-term task of assisting in the economic development of its member countries.[28]

Lending for development purposes was extended by 1949 to some other countries outside the western hemisphere. India received its first loan in 1949 for financing the purchase of locomotives and spare parts for its railway system.[29] Two more loans to India followed. The second loan was made in late 1949 for financing the import of equipment for reclamation of agricultural land, and the

---

[23] *See id.*, at 19.

[24] *See id.*, at 24.

[25] *See id.*, at 24.

[26] *See supra*, n. 17, at 8. An initial sum of $5 billion was made available under the Marshall Plan for the purpose of the economic recovery program in Europe, for a period of over four years.

[27] For a further discussion of the impact of the Marshall Plan on Bank operations, *see supra* n. 20, at 14–15.

[28] *See* International Bank for Reconstruction and Development, *Tenth Annual Report*, 1954–55, at 30.

[29] *See supra* n. 20, at 29–31.

third loan was made in 1950 for financing part of the cost of developing the resources of the Damodar Valley.[30] Although the first two loans to India were in the nature of program lending, similar to earlier Bank loans, the third loan was more in the nature of project lending since it was earmarked for specified investment purposes, in a defined geographical area. Shedding more light on the Damodar Valley project, the Bank's *Sixth Annual Report* indicated that the project was progressing satisfactorily. The Report went on to state that although the first thermal unit would be ready by the end of 1952, "its operation through the subsequent dry season, however, will depend largely on the amount of cooling water provided by the Konar Dam, and completion of the dam will be delayed because of changes in design."[31]

This approach to development lending was gradually expanded. In 1950 the Bank extended its first loan to Iraq for the construction of the Wadi Tharthar flood control project on the Tigris River.[32] The project was designed to prevent recurring floods in both agricultural lands and residential areas. The project was "part of a more comprehensive plan, which will eventually provide, in addition, for water storage, regulation of water supply and irrigation. The initial project calls for the construction of a dam across the Tigris River at a point about 50 miles above Baghdad, which will direct excess flood waters into an uninhabited and barren depression, known as Wadi Tharthar, situated between the Euphrates and the Tigris rivers northwest of Baghdad."[33]

Hence, the emphasis on lending to the power and irrigation sectors gradually led the Bank into the delicate and complex area of rivers shared by two or more countries, as will be discussed in the next Part of this Chapter.

1.2   Initial Bank Operations on International Waterways

As discussed above, the Bank started providing loans to its member countries in about a year after it formally commenced operations. The first four loans were intended to assist its European members in the reconstruction and development of their economies. Some of those loans, as well as the reconstruction loans granted later, came close to being granted "for general purposes to be determined

---

[30] *See* International Bank for Reconstruction and Development, *Fifth Annual Report*, 1949–50, at 27.

[31] *See* International Bank for Reconstruction and Development, *Sixth Annual Report*, 1950–51, at 21.

[32] *See supra* n. 30, at 30.

[33] *See id.*

by the country itself."[34] However, the Bank moved gradually toward financing equipment and materials for specific sectors of its borrowing countries. This was exemplified by the Chilean loans, parts of which financed hydroelectric plants, related transmission lines and installation of additional generating units, as well as the provision of pumping equipment for irrigation. The equipment and the related services were not intended for use in a defined geographical area, and as such could be used across the entire country. Because the goods and equipment were to be used for hydroelectric power and irrigation, they could be used for projects on rivers that run exclusively in Chile, as well as on rivers that Chile shares with some of its neighbors. Loans for similar purposes—expansion of hydroelectric generating stations—were extended to Mexico and Brazil. Since the equipment purchased through those loans were to be used all over both countries, and since those countries share a number of rivers with their neighbors, it could be concluded that the Bank had started, as early as the late 1940s to finance projects on international rivers.[35]

The Bank paid particular attention in its early years to hydroelectric power projects. In 1954 the Bank noted that "A dearth of basic services, and particularly of electric power and transportation, has been the major physical obstacle to increasing production and raising living standards in the less developed countries. Lack of these services put severe limits on productivity, on income and on willingness to invest. Deficiencies of electric power supply were conspicuously handicapping industrial growth."[36] Moreover, the Bank's emphasis on lending to power projects at that time was necessitated by "the very rapid growth of power demand in recent decades. In the world as a whole, demand has increased since the war at an average of nearly 10 percent a year, and this expansion, which reflects the raising of living standards as well as the continued growth of economic activity is expected to continue."[37] This emphasis on lending for hydroelectric power led, by necessity, to financing of projects on rivers, whether national or international.

However, at the time such loans were processed in the late 1940s and early 1950s, not much attention was paid to whether the river was a wholly national

---

[34] *See* Mason & Asher, *supra* n. 2, at 271.

[35] An international river is one either flowing through the territory of two or more countries (also referred to as a successive river), or one separating the territory of two countries from one another (also known as a boundary or contiguous river.) The definition also includes any tributary to such a river. *See The Law of International Drainage Basins*, 16 (A. H. Garretson *et al.*, eds., Oceana Publications 1967). The terms "international," "shared" and "transboundary" are used inter-changeably throughout this Book.

[36] *See supra* n. 28, at 30.

[37] *See* International Bank for Reconstruction and Development, *Eleventh Annual Report*, 1955–56, at 8.

river, or whether such a river was shared with other countries. This became clearer when the Bank extended its second loan to India in 1950 for developing the resources of the Damodar Valley. The project included the construction of the Konar Dam over the Damodar River. The Damodar River is a tributary of the Ganges River which is an international river shared by India, Nepal, Bangladesh (at that time was East Pakistan) and China.[38] Similarly, the Wadi Tharthar project in Iraq which was approved in 1950 also involved the construction of a dam over the Tigris River which Iraq shares with Turkey and Syria.[39] In Africa, the Bank extended in 1952 its first loan to the Belgian Congo (present-day Democratic Republic of Congo, or DRC) for assisting in the implementation of the country's ten-year development plan.[40] The activities financed by the Bank loan included improving navigation on the Congo River through the installation of navigation lights and the provision of additional tugboats and barges.[41]

These projects raised no special concerns. They were dealt with in the Bank reports in a manner that was not in any way different from other projects on national rivers, such as those in Thailand, Turkey and Brazil. In Thailand, the Bank financed a power development project over the Chao Phya River,[42] and in Turkey the construction of a dam over the Seyhan River to help control floods and provide water for irrigation. In Brazil the Bank extended a loan in 1950 to *Companhia Hirdo Eletricado São Francisco* for power development.[43] The

---

[38] For more information on the Damodar River, *see* B. R. Chauhan, *Settlement of International and Inter-state Water Disputes in India*, 133 (India Law Institute 1992).

[39] For more details on the Tigris and Euphrates rivers, *see* Ayşegül Kibaroğlu, *Building a Regime for the Waters of the Euphrates-Tigris River Basin* (Kluwer Law International 2002). It should be added that the Diyala River which originates in Iran, flows into Iraq where it joins the Tigris, thus adding Iran to the riparian states.

[40] *See* International Bank for Reconstruction and Development, *Seventh Annual Report*, 1951–52, at 14. The Report described the loan as the first loan "made by the Bank to an overseas territory of a member country." The member country in this case was Belgium. *See id.*, at 14.

[41] *See supra* n. 28, at 17. The Congo River is shared by Angola, Burundi, Cameroon, the Central African Republic, the Democratic Republic of Congo, the Republic of Congo, Rwanda, Tanzania, and Zambia. For more on the Congo River, *see* Piet Heyns, *Water Resources Management in Southern Africa*, in *International Waters in Southern Africa* 28–29 (Mikiyasu Nakayama, ed., United Nations University Press 2003). At the time the Loan was made, these countries were under the control either of Great Britain, France, Belgium or Portugal.

[42] *See supra* n. 31, at 24. The Project included the construction of a barrage to partially retain the waters of the Chao Phya River, with housing for the future installation of power-generating equipment.

[43] *See* International Bank for Reconstruction and Development, *Ninth Annual Report*, 1953–54, at 28–29.

project involved construction of a power station over the São Francisco River which runs exclusively within Brazil. In 1953 the Bank extended another loan to Brazil for expansion of hydroelectric power.[44] This project involved the construction of a dam at Itutinga Falls on the Rio Grande which is a tributary of the São Francisco River.

One main reason why the Bank did not pay particular attention to the fact that some of its financed projects were on international rivers was because, at that time, there were no apparent disputes over the sharing of those particular rivers among the different riparians. Moreover, the Bank-financed projects were not expected to raise concerns with any of the other countries as they were not expected to have adverse effects on other riparians. These factors might have caused the issue of the rivers being shared with other countries to go unnoticed, or at best, to raise no concerns. The projects were processed without any questions being asked regarding the issue of the transboundary nature of those rivers.

However, soon afterward, it appeared to the Bank that some of the proposed power and irrigation projects were on rivers where there were existing or emerging disputes between some of the riparians, or the projects were expected to raise concerns from other riparians because of their likely adverse effects. In 1950, the Bank began discussions with Syria on the possibility of the Bank financing various projects in Syria, including projects for flood control, irrigation and drainage.[45] One of those projects, Draining of the Ghab Project (the Ghab Project), sought to drain and reclaim the Ghab swamplands in Syria.[46] The project was located in the valley of the Orontes (or the Asi) River which Syria shares with Lebanon and Turkey.[47] The other project discussed with Syria was the Youssef Pasha Multipurpose Project (Youssef Pasha Project) which was an irrigation and electric power project on the Euphrates River.[48] Syria shares the Euphrates River with Iraq and Turkey. The Bank also held discussions with Egypt on similar projects: improvements to, and the expansion of, irrigation in Upper and Lower Egypt, and hydroelectric power combined with industrial development at Aswan.[49] Those projects involved the Nile

---

[44] *See id.*, at 28–29.

[45] *See supra* n. 31, at 23.

[46] *See supra* n. 43, at 18.

[47] For more details on the Orontes River, *see* Greg Shapland, *Rivers of Discord—International Water Disputes in the Middle East*, 144 (Hurst & Company 1997).

[48] *See* International Bank for Reconstruction and Development, *Eighth Annual Report*, 1952–53, at 20.

[49] *See supra* n. 31, at 17.

River that Egypt shares with a number of other countries, and included the Aswan High Dam Project which was expected to flood large areas in northern Sudan.[50] Similarly, discussions commenced in 1949 between the Bank and India on the financing of the Bhakra-Nagal Dam Project, and in 1950 between the Bank and Pakistan on the financing of the Lower Sind Barrage Project, both on the Indus River. By that time the dispute over the Indus River Basin between India and Pakistan had already erupted and escalated, following the partitioning of the Sub-continent.

Confronted with these complex situations, the Bank started its internal deliberations on how to handle these projects, and how to involve the other riparians and ensure that their interests were not adversely affected by the proposed projects. The difficulty of the situation was further compounded by the paucity of international law rules in the field of shared rivers. As will be discussed in the next Part of this Chapter, at that time there were no established or widely accepted principles of international law governing the uses and sharing of international rivers.

1.3   International Water Law in the 1950s

It should not come as a surprise that the Bank did not pay special attention at the very beginning of its operations to the projects it was financing that turned out to be situated on, or affecting, international rivers. Nor should it be surprising that the Bank was not able to immediately map a policy for such projects when the fact that those rivers are shared by two or more states became apparent. At that time, the late 1940s and early to mid-1950s, international water law, particularly with regard to non-navigational uses of international watercourses, was at an infancy stage, and there were no established principles to guide the Bank's work in this field.

Europe started its attempts to deal with shared rivers by addressing navigational issues. Such attempts started in the early years of the nineteenth century. At that time, navigation was, by and large, the single and dominant user of rivers. Non-navigational uses such as irrigation, hydropower and industry were in the initial stages of development.[51] The early efforts of Europe to deal with

---

[50] *See supra* n. 48, at 19.

[51] It should be added that, in addition to the navigational and non-navigational uses, international rivers and lakes may also be used to delimit boundaries between countries. Although boundaries, as a general rule, are established by treaties, the interpretation of treaties as to where the boundaries actually fall across the river or the lake often has to take into account the particular characteristics of the international river or lake in question. For

shared rivers resulted in the adoption of the Final Act of the Congress of Vienna on June 9, 1815. The Act established the principle of freedom of navigation on shared rivers, for all riparians, on a reciprocal basis, as well as priority of navigation over other uses.[52] The trend toward freedom and priority of navigation established by this Act continued to prevail, and was confirmed in 1885 by the General Act of the Congress of Berlin with regard to the Congo and Niger rivers,[53] thus extending the freedom of navigation outside Europe, as well as to non-riparian states. The 1919 Peace Treaty of Versailles continued the liberalization trend in navigation by opening all the navigable rivers in Europe to all the European countries. This trend was confirmed 10 years later by the Permanent Court of International Justice (PCIJ) in the *River Oder* Case, where the PCIJ ruled that the

---

a general discussion of international rivers as boundaries, *see* R. R. Baxter, *The Law of International Waterways* (Harvard University Press 1964); *see* also Salman M. A. Salman, *International Rivers as Boundaries—The Dispute over the Kasikili/Sedudu Island and the Decision of the International Court of Justice*, 25 Water International 580 (2000); *see* also *Case Concerning Kasikili/Sedudu Island (Botswana/Namibia)*, I.C.J. Reports 1999. Subsequent to the Kasikili/Sedudu case, the International Court of Justice (ICJ) has decided two other cases involving boundary rivers. The first dispute arose between Cameroon and Nigeria over their maritime boundaries around the Bakasi Island, and their land and water boundaries around the area of Lake Chad Basin. The ICJ issued its decision on this dispute in 2002 [*Land and Maritime Boundary between Cameroon and Nigeria* (Cameroon v. Nigeria; Equatorial Guinea intervening), Judgment, I.C.J. Reports 2002]. The second one arose between Benin and Niger over their borders across the Niger River sector, and on the ownership of islands in the River. A ruling by the ICJ was issued in 2005 [*Frontier Dispute (Benin/Niger)*, Judgment, I.C.J. Reports 2005]. In September 2005, the ICJ registered a fourth similar dispute, this time from Central America, *Dispute regarding Navigational and Related Rights* (Costa Rica v. Nicaragua). Costa Rica brought the case against Nicaragua over the San Juan River, which forms the borders between the two countries. Costa Rica does not seem to dispute the ownership of the entire San Juan River by Nicaragua under the 1858 Treaty of Limits. Rather, Costa Rica is claiming that it has navigational rights under said Treaty, and that Nicaragua is imposing a number of restrictions on its exercise of such navigational rights. This case still awaits a decision by the ICJ. For each of those cases, *see*: http://www.icj-cij.org/docket/index.php?p1=3&p2=2. Although international water law covers navigational and non-navigational uses, as well as boundary issues across rivers and lakes, the emphasis has gradually shifted in the last century to the non-navigational uses.

[52] For the text of the Final Act of the Congress of Vienna, s*ee Fontes Historiale Iuris Gentium, (Sources Relating to the History of the Law of Nations)*, 455 (Wilhelm G. Grewe, ed., vol. 1 1992). *See generally*, Karl von Martens & Ferdinand de Cussey, *Recueil Manuel et Pratique de Traités, Conventions et Autres Actes Diplomatiques* (1935).

[53] *See* 3 Am. J. Int'l L., (Supp), at 7 (1909). The purpose of the expansion of the concept of freedom of navigation to the Congo and Niger rivers was to facilitate the conquest of Africa by the European colonial powers. For more details *see* Béla Vitanyi, *The International Regime of River Navigation*, 98 (Kluwer Law International 1979). *See also* Baxter, *supra* n. 51, at 149.

jurisdiction of the International Commission for the Oder River extended to certain tributaries of that river.[54]

The changing economic circumstances, however, gradually eroded the primacy of navigation established at the beginning of the nineteenth century. The industrial revolution and the steady increase in population necessitated other uses for rivers. The industrial revolution brought about new modes of transportation and carriage of goods that reduced the importance of rivers as international highways. With the beginning of the twentieth century non-navigational uses of international rivers started evolving and competing with navigational uses. In 1921 the Convention and Statute on the Regime of Navigable Waterways of International Concern, which is also known as the Barcelona Convention, was adopted under the auspices of the League of Nations.[55] Although the Convention confirmed the principle of freedom of navigation, it did recognize non-navigational uses as well. Article 10(6) conceded the right of a riparian state, in exceptional circumstances, to close the waterway, wholly or in part, to navigation if navigation was of little importance to it, and if the state could justify its action on the ground of economic interest clearly greater than that of navigation. This started a process of decline of both the primacy and freedom of navigation. After the Second World War and the division of Europe into east and west camps, freedom of navigation was gradually restricted only to the riparian countries of the particular shared river on a reciprocal basis. This situation has continued to prevail, and represents contemporary customary

---

[54] *See Case Relating to the Territorial Jurisdiction of the International Commission of the River Oder* (United Kingdom, Czechoslovakia, Denmark, France, Germany and Sweden v. Poland), Judgment No. 16 (PCIJ., Ser. A, No. 23, 1929). The issue before the PCIJ was whether the jurisdiction of the International Commission of the Oder River extended, under the provisions of the Treaty of Versailles (1919), to sections of the tributaries of the Oder River, namely the Warthe and Netze, which were situated in Poland. Whereas Poland maintained that the jurisdiction of the Oder Commission excluded sections of those tributaries in Polish territory, the other six countries felt that the navigable portions of those tributaries should not be excluded from the Commission's jurisdiction. The Court ruled that the jurisdiction of the Commission extended to certain tributaries of the Oder River situated within Poland. Along these lines of liberalization, the PCIJ in the *Oscar Chinn* case (Great Britain v. Belgium), Judgment No. 23 (PCIJ., Ser. A./B., No. 63, 1934) stated, five years after its decision in the River Oder case, that ". . . freedom of navigation implies, as far as the business side of maritime or fluvial transport is concerned, freedom of commerce also." *See id.*, at 83. As we will see later, the Oder River is also known as the Odra River. *See infra* n. 420.

[55] 7 L.N.T.S. 35 (1921). The Convention was signed on April 20, 1921 in Barcelona; *see* also *International Environmental Law: Multilateral Treaties* (W. E. Burhenne, ed., vol.1, Kluwer Law International 1996). The Convention included procedural aspects such as signature, accession and ratification, while the Statute laid down detailed rules regarding the navigable uses of international waterways.

international law in this field,[56] as reflected in a number of regional and multi-lateral watercourses agreements.[57]

It should be noted, however, that the decline in the primacy and freedom of navigation was not accompanied by the emergence of clear and binding rules regulating the non-navigational uses of international rivers. There were some principles, rules, conventions and judicial and arbitral decisions governing some aspects of the non-navigational uses of international rivers. However, some of those principles conflict with some others, and the status of most of the rules is unclear. The issues addressed by the relevant conventions were limited and not many countries have signed and ratified them.

The four principles that addressed issues related to the non-navigational uses of international rivers were based, to a large extent, on state practice and the work of scholars and experts in the field. The first of those principles, absolute territorial sovereignty, was, from the start, the most controversial. According to this principle, also known as the Harmon Doctrine, a state is free to dispose, within its territory, of the waters of an international river flowing within its territory in any manner it deems fit, without concern for the harm or adverse impact that such use may cause to other riparian states.[58] In essence, the principle

---

[56] *See* Lucius Caflisch, *Regulation of the Uses of International Watercourses*, in *International Watercourses—Enhancing Cooperation and Managing Conflict*, World Bank Technical Paper No. 414, at 7 (Salman M. A. Salman & Laurence Boisson de Chazournes, eds., The World Bank 1998).

[57] By way of examples *see* Treaty for Amazonian Cooperation concluded by the eight riparian states of the Amazon River (Bolivia, Brazil, Colombia, Ecuador, Guyana, Peru, Suriname and Venezuela) in Brasilia on July 3, 1978. Article 3 of the Treaty states that ". . . the Contracting Parties mutually guarantee on a reciprocal basis that there shall be complete freedom of commercial navigation on the Amazonian rivers. . . ." For the text of the Treaty, *see* 17 I.L.M. 1045 (1978). The Revised Protocol on Shared Watercourses in the Southern African Development Community (SADC), (hereinafter the "SADC Protocol") concluded on August 7, 2000, by the members of SADC (*see infra* n. 248) makes a similar reference. Article 3(2) of the SADC Protocol states that "The utilisation of shared watercourses within the SADC Region shall be open to each Watercourse State, in respect of the watercourses within its territory and without prejudice to its sovereign rights, in accordance with the principles contained in this Protocol. The utilisation of the resources of the watercourses shall include agricultural, domestic, industrial, navigational and environmental uses." For the SADC Protocol *see* 40 I.L.M. 321 (2001).

[58] This principle is based on the opinion given by the Attorney General of the United States, Mr. Judson Harmon, in 1895 regarding a dispute with Mexico over the utilization of the waters of the Rio Grande. The opinion concluded that there was no settled and recognized right "by which it could be held that the diversion of the waters of an international boundary stream for the purpose of irrigating lands on one side of the boundary and which would have the effect to deprive lands on the other side of the boundary of water for irrigation purposes would be a violation of any established principle of international law." *See* 21 Op. Att'y Gen. 274, 283 (1895).

of absolute territorial sovereignty attests that there is no specific international law on the subject.

The second principle pertains to absolute territorial integrity. It establishes the right of a riparian state to demand continuation of the natural flow of an international river into its territory from the upper riparian or riparians, but imposes a duty on that state not to restrict such natural flow of waters to other lower riparians. At most, this principle tolerates only minimal uses by an upstream state, and in that respect it has similarities with the common law doctrine of riparian rights.[59] As such, this principle is the exact opposite of the principle of absolute territorial sovereignty as it is intended to favor downstream riparians, often by protecting existing uses or prior appropriations. The principle is supposed to protect lower riparians from any harm or injury that may be caused by the uses of the waters of the international river by the upper riparian. This contrasts with the principle of absolute territorial sovereignty where the state, often the upper riparian, is free to dispose of the waters of the international river the way it deems fit. Those two principles were rejected by subsequent state practice, including that of the United States of America itself, as manifested in the bulk of agreements on shared watercourses.[60]

---

[59] Under the common law doctrine of riparian rights, as originally conceived, the upper riparian land owner should allow the water to flow down its natural course without interfering with its quantity or quality. However, the principle of riparian rights gradually developed to allow each riparian a reasonable use of the available amount of water. For a discussion of the doctrine of riparian rights, *see* Ludwik A. Teclaff, *Water Law in Historical Perspective* 6–20 (William S. Hein & Co. 1985). *See* also, David H. Getches, *Water Law in a Nutshell* 15–55 (4th ed., West Publishing Company 1997).

[60] As an example, the 1909 Treaty between the United States and Great Britain Relating to the Boundary Waters between the United States and Canada recognizes the rights of each of those countries over their shared boundary rivers. *See Treaties and Other International Agreements of the United States of America* 1776–1949, 319 (1968–1976), (Charles Bevans, ed.). This Treaty was concluded less than 15 years after Harmon issued his legal opinion. Furthermore, the 1944 Treaty between the United States of America and Mexico Relating to the Utilization of the Waters of the Colorado and Tijuana Rivers and of Rio Grande (Rio Bravo) from Fort Quitman, Texas, to the Gulf of Mexico (*see* Bevans, *id.*, at 1166), follows primarily the principle of equitable and reasonable utilization by allotting to each of the two countries certain specified amounts of the waters of those rivers. As early as the 1930s, a leading authority on international water law described the Harmon Doctrine as "radically unsound." After stating that the principle of absolute territorial sovereignty would give the upper riparian the right to exhaust the whole waters of the river irrespective of the injury it would cause to the lower riparian, and that it would equally entitle the lower riparian to flood the lands of the upper riparian, Herbert Smith concluded "It seems obvious that a principle which leads to those consequences must be radically unsound." *See* Herbert Arthur Smith, *The Economic Uses of International Rivers*, 8 (P. S. King & Son, Ltd. 1931). For an analysis of the Harmon Doctrine, *see* also, Stephen McCaffrey, *The Harmon Doctrine One Hundred Years Later: Buried, Not Praised*, 36 Nat. Resources J. 549 (1996).

The third principle is that of limited territorial sovereignty. It restricts both the principles of absolute territorial sovereignty and absolute territorial integrity by asserting that every riparian state has a right to use the waters of an international river, but is under a corresponding duty to ensure that such use does not significantly harm the other riparians. In essence, this principle establishes the equality of all riparian states, and confirms the right of each of them over the shared river through restricting and reconciling the first two principles.

The fourth principle is the community of co-riparian states in the waters of an international river.[61] According to this principle, the entire river basin is an economic unit, and the rights over the waters of the entire river are vested in the collective body of the riparian states, or divided among them either by agreement or according to proportionality.[62] However, lack of trust among most riparian states, resulting in competing, and sometimes conflicting, planning activities undermines this idealistic approach towards cooperative and integrated development of the shared river basin.

This leaves the third principle of limited territorial sovereignty as the most widely accepted principle. The subsequent developments in international water law would evolve around this principle, since it combines both the concept of equitable and reasonable utilization, and the obligation not to cause harm, as will be discussed in the course of this Book.

In addition to these principles, reference should also be made to the work of two scholarly non-official bodies in the field of international water law. The first of these bodies is the Institute of International Law (IIL, known more commonly by its French name, *Institut de Droit International*).[63] The IIL issued its

---

[61] For discussion of those four principles *see* Jerome Lipper, *Equitable Utilization* in Garretson *et al* (eds.), *supra* n. 35, at 18. *See* also F. J. Berber, *Rivers in International Law*, 11 (Stevens & Sons 1959), and Salman M. A. Salman & Kishor Uprety, *Conflict and Cooperation on South Asia's International Rivers—A Legal Perspective*, (Kluwer Law International 2002) at 11.

[62] In the *River Oder* case, the PCIJ referred to "the principles governing international fluvial law in general." After discussing the principle of community of interest of riparian states, the Court concluded that "This community of interest in a navigable river becomes the basis of a common legal right, the essential features of which are the perfect equality of all riparian States in the user of the whole of the course of the river and the exclusion of any preferential privileges of any one riparian State in relation to the others." *See Case Relating to the Territorial Jurisdiction of the International Commission of the River Oder, supra* n. 54). As will be discussed later (*infra* Chapter 4, Part 4.5), the ICJ confirmed this principle in 1997, almost 70 years later, in the *Gabčíkovo-Nagymaros* case, *infra* n. 326.

[63] The Institute of International Law (IIL) was established in 1873 and consists of selected members. The IIL "is a purely scientific and private association, without official character, whose objective is to promote the progress of international law by: formulating

first resolution on international rivers entitled "International Regulations Regarding the Use of International Watercourses for Purposes Other than Navigation" (known also as the "Madrid Declaration") in 1911.[64] The Declaration makes a distinction between streams that form boundaries between two states, and streams that traverse successively through two or more states.[65] For frontier streams, the first article of the Declaration states that: ". . . neither of these [riparian] States may, without the consent of the other, and without special and valid legal title, make or allow individuals, corporations, etc. to make alterations therein detrimental to the bank of the other State. On the other hand, neither State may, on its own territory, utilize or allow the utilization of the water in such a way as seriously to interfere with its utilization by the other State, or by individuals, corporations, etc., thereof."[66]

The second article of the Madrid Declaration deals with successive streams and obliges the riparian states not to change the point where such a stream crosses the frontiers of such states, by means of establishments of one of the states without the consent of the other riparians. It also forbids all alterations injurious to water, and the emptying therein of injurious matter such as from factories. The Declaration also prohibits establishments such as factories from taking so much water as to seriously modify such stream. It also prohibits construction that would subject other states to inundation.

As such, the Declaration codifies the principle of not causing appreciable harm to the other riparians. It should be noted that under this principle a state is entitled to use the flow of an international watercourse in its territory but only in such manner as not to cause appreciable harm to another riparian.[67] Although the

---

general principles; cooperating in codification; seeking official acceptance of principles in harmony with the needs of modern society; contributing to the maintenance of peace or to the observance of the laws of war; proffering needed judicial advice in controversial or doubtful cases; and contributing, through publications, education of the public, and any other means, to the success of the principles of justice and humanity which should govern international relations." For this statement and for more information on the IIL, *see:* http://nobelprize.org/nobel_prizes/peace/laureates/1904/international-law-history.html. As will be discussed in this Book, the IIL has adopted a number of resolutions on international water law.

[64] The Declaration was adopted by the IIL at its session at Madrid, April 20, 1911; 24 Annuaire de l'Institut de Droit International 365 (1911).

[65] It should be noted that international law, in terms of rights and duties, does not draw any legal distinction between boundary rivers and successive rivers. The same rules of international law apply to both types of rivers. *See River Oder Case, supra* n. 54, at 27.

[66] *See* the Madrid Declaration, Article 1. The same Article extends the application of this rule to "a lake lying between the territories of more than two States." Perhaps the intention was "two or more states," rather than "more than two states."

[67] The no harm rule, *sic utere tuo ut alienum non laedas* (so use your own property as not to injure your neighbor) covers, under international law, a whole range of neighborly relations,

Madrid Declaration embodies the principle of limited territorial sovereignty, it goes on to establish absolute prohibition against activities that would result in injury to the other riparians, without their consent. Moreover, the Declaration rejects the principle of absolute territorial sovereignty by imposing certain restrictions on each riparian state, including the requirement of obtaining the consent of the other riparians. Consequently, the Declaration went to the other extreme by requiring consent of other riparians for any works that seriously interfere with their utilization of the shared watercourse.[68]

It was not until 1961, 50 years after the Madrid Declaration, that the IIL issued its second resolution in this area, addressing the requirement of consent under the Declaration. That resolution, known as the Salzburg Resolution,[69] will be discussed in the next Chapter.

The other scholarly non-official body whose work has an important impact on the developments of international water law is the International Law Association (ILA).[70] Unlike the IIL, the ILA did not start working on international water law until 1954, following the establishment of the "Rivers Committee," which later was renamed the "Committee on the Uses of the Waters of International Rivers."[71] In 1956 the ILA issued its first set of principles entitled "A Statement of Principles upon which to Base Rules of Law Concerning the Uses of International

---

including issues pertaining to the protection of the environment, as was noted in the *Trail Smelter arbitration, infra* n. 79.

[68] It is worth noting that the Madrid Declaration used the term "international watercourses" rather than "international rivers." As will be discussed later, the term "international watercourses" will surface later and prevail over other terms in this field.

[69] Resolution on the Utilization of Non-Maritime International Waters (Except for Navigation), adopted by the Institute of International Law at its Session at Salzburg, September 11, 1961, 49 Annuaire de l'Institut de Droit International 370 (1961); *see* also, 56 Am. J. Int'l L. 737 (1962).

[70] Like the IIL, the International Law Association (ILA) was founded in 1873. However, membership of the ILA is open to all international lawyers and, unlike the IIL, is not by invitation. The objectives of the ILA, under its Constitution, are "the study, clarification and development of international law, both public and private, and the furtherance of international understanding and respect for international law." The ILA has consultative status, as an international non-governmental organization (NGO), with a number of the United Nations specialized agencies. For more information on the ILA, *see*: http://www.ila-hq.org/. *See* also, Slavko Bogdanovic, *International Law of Water Resources—Contribution of the International Law Association* (Kluwer Law International 2001).

[71] For a detailed account of the work of the ILA in this field, *see* Charles Bourne, *The International Law Association's Contribution to International Water Resources Law*, 36 Nat. Resources J. 155 (1996). *See* also Charles Bourne, *International Water Law, Selected Writings of Professor Charles Bourne*, Chapter 3, at 83 (Patricia Wouters, ed., Kluwer Law International 1997). *See* also Bogdanovic, *id.* "The Committee on the Uses of the Waters of International Rivers" was also known as the "Rivers Committee."

Rivers," also known as the "Dubrovnik Statement."[72] The Statement of Principles
confirms the sovereign control each state has over the international river within
its own boundaries but requires that state to exercise such control with due
consideration for its effects upon other riparian states.[73] As such, the ILA rejected
the principle of absolute territorial sovereignty, as well as the principle of
absolute territorial integrity, and recognized clearly the right of all riparian states
over the shared river. Unlike the Madrid Declaration, the Dubrovnik Statement
did not require the consent of the other riparians as a prerequisite for carrying out
any works on the shared river.

Thus, by the time the Bank started financing projects on international rivers in
the late 1940s and early 1950s, the work of the IIL and ILA was in its infancy
stage, and neither of them had developed any clear principles in the field of international water law. The Madrid Declaration had actually imposed a very strict and
unreasonable requirement regarding the consent of the other riparians—a
requirement that clearly did not represent customary international law. On the
other hand, the ILA had barely started paying attention to international rivers, and
its Committee on the Uses of the Waters of International Rivers was only established in 1954. Although the Dubrovnik Statement acknowledges the rights of all
the riparian states over the shared rivers, it did not provide any guidance as to how
such rights could be determined and exercised.

Two conventions relevant to the non-navigational uses of rivers were adopted
under the auspices of the League of Nations, namely, the Barcelona Convention,[74]
and the Geneva Convention (or the General Convention Relating to the Development of Hydraulic Power Affecting More Than One State).[75] The Barcelona Convention was adopted in 1921, 10 years after the Madrid Declaration was issued,
while the Geneva Convention was concluded in 1923, two years after the
Barcelona Convention. As discussed earlier, the Barcelona Convention has little
relevance to non-navigational uses of rivers. It simply recognizes non-navigational
uses of international rivers, and allows a riparian state to close the waterway,
wholly or in part, to navigation if navigation was of little importance to it, and if
the state could justify its action on the ground of economic interest clearly greater
than that of navigation.

---

[72] ILA, Report of the Forty-Seventh Conference (Dubrovnik 1956), at 241 (hereinafter the
"Dubrovnik Statement"). The intention of the Rivers Committee was not to state rules of
law, but only to lay down principles on which rules of law could be formulated. *See*
Bourne, *International Water Law, supra* n. 71, at 159–60.

[73] *See* paragraph 3 of the Dubrovnik Statement, *id.*, at 241–42.

[74] For the Barcelona Convention, *see supra* n. 55.

[75] 36 L.N.T.S. 75 (1923). The Convention was signed in Geneva on December 9, 1923.

The Geneva Convention basically dealt with development of hydropower on shared rivers, and as such, could be regarded as the first international instrument to attempt addressing this subject, albeit in a very limited way. The Convention addressed the right of any riparian state to carry out in its territory any operations for the development of hydraulic power that it may consider desirable, subject to "the limits of international law."[76] The Geneva Convention calls for joint investigation by the Contracting States for the reasonable development of hydraulic power, and obliges the parties to pay due regard to any works already existing, under construction, or projected. If such development involves the use of the territory of another state, or may cause prejudice to another state, those states shall enter into negotiations with a view to conclude an agreement.

It should be noted, however, that those two conventions actually had little significance with regard to the development of principles for non-navigational uses of rivers.[77] In reality, they did not address in any way how the rights and obligations of the riparian states are to be dealt with. The contribution of the Barcelona Convention was that it paved the way for non-navigational uses of international rivers. With regard to the Geneva Convention, ". . . the seemingly absolute prohibition of the Madrid Declaration against upper riparians undertaking construction that might alter the regime of the waters was superseded at Geneva by the principle of reasonableness, and the necessity of negotiations."[78]

Similarly, in the early 1950s when the Bank started paying attention to projects on international rivers, there were few judicial or arbitral decisions that could guide its work in this area. In addition to the *River Oder* case discussed earlier, three other cases that were interpreted to have relevance to international rivers are worth noting. These cases are the *Trail Smelter* and the *Lake Lanoux* which were decided through arbitration, and the *Corfu Channel* case which was decided by the International Court of Justice (ICJ). In 1941, the Arbitration Tribunal in the *Trail Smelter* case concluded that ". . . no state has the right to use or permit the use of its territory in such a manner as to cause injury by fumes in or to the territory of another or the properties or persons therein, when the case is of serious consequence and the injury is established by clear and convincing

---

[76] *See* Article 1 of the Geneva Convention. The Convention does not indicate what those limits of international law are.

[77] Some jurists, such as Berber, do not regard the Barcelona Convention ". . . as the expression of prevailing state practice, as in the first 15 years after its conclusion it was ratified by only 20 states. These states, moreover, consisted almost entirely of states in whose territory there were no rivers to which the Convention was applicable." *See* Berber, *supra* n. 61, at 123.

[78] *See* ILA, *Principles of Law and Recommendations on the Uses of International Rivers*, at 56 (London, 1959).

evidence."[79] Applied to international rivers, the conclusion of the Arbitral Court seems to rephrase part of the principles enunciated by the IIL in the Madrid Declaration by prohibiting acts that would have adverse effects by one riparian state on another.[80]

Along those lines, the ICJ confirmed in the *Corfu Channel* case in 1949 the principle of state responsibility for acts contrary to international law that occur within the territory of a state and result in injury to another party.[81] The *Lake Lanoux* arbitration dealt with a dispute between France and Spain over Lake Lanoux which the two countries share. The tribunal ruled that although France was entitled to use the waters of Lake Lanoux, it could not ignore Spain's interests, and that Spain was entitled to demand that Spanish rights be respected and its interests be taken into account.[82] Those cases, particularly the *Trail Smelter* and *Lake Lanoux*, have had considerable influence on the development of international water law.[83]

---

[79] *See Trail Smelter Arbitration* (United States v. Canada), 3 R.I.A.A. 1911, 1965 (1941). In this case smelting operations in British Columbia in Canada resulted in sulphur dioxide fumes being emitted into the air, causing damage to United States citizens across the border in Washington State. The matter was referred to the International Joint Commission established under the 1909 Boundary Waters Treaty (*supra*, n. 60). The recommendations of the Commission that included, *inter alia*, assessment of the damage up to 1931 at $350,000, were rejected by the United States. As a result, the parties concluded the "Convention for the Settlement of Difficulties Arising from Operation of the Smelter Trail" between Great Britain and the United States on April 15, 1935. The Convention provided for the establishment of a three-member tribunal which dealt with the matter and issued its decision in 1941.

[80] For the Madrid Declaration *see supra* n. 64.

[81] *See Corfu Channel* case, (U.K. v. Albania), I.C.J. Reports 4, 1949, at 3. In this case, a number of British officers lost their lives in 1946 when their warship struck mines in the Corfu Channel in the territorial waters of Albania. The International Court of Justice, confirming the principle of harmless use of territory, concluded that under international law, every state is under obligation not to knowingly allow its territory to be used for acts contrary to the rights of other states. As such, Albania was responsible for the explosion of such mines and for the resulting damage and loss of lives.

[82] *See Lake Lanoux Arbitration* (France v. Spain), 24 I.L.R. 101 (1957); and also 53 Am. J. Int'l L. 156 (1959). The arbitration dealt with the proposed use by the French government of the waters of Lake Lanoux for carrying out certain hydroelectric works. The Lake is fed by some streams originating and flowing through French territory. Its waters emerge by the Font-Vive stream that forms one of the headwaters of the River Carol, which flows into Spain, where it joins the River Segre. Spain opposed the use of the waters of Lake Lanoux by France, claiming adverse effects as a result. While the Tribunal decided that the works did not constitute infringement of Spain's rights, it also ruled that France was entitled to use the waters of Lake Lanoux but could not ignore Spain's interests, and that Spain was entitled to demand that Spanish rights be respected, and its interests be taken into account.

[83] In addition to these cases, a reference should be made to two other cases decided by the Permanent Court of International Justice (PCIJ). The first case is the Oder River case to

Accordingly, when the Bank was confronted in the early 1950s with the complex issues of projects on international waterways, there were no established rules and principles of international water law to guide its work in the field of non-navigational uses. Some of the principles which the academic circles were deliberating contradicted each other, and did not have wide acceptance. The work of the IIL and that of the ILA on shared rivers at the beginning of the 1950s was meager. The IIL requirement of consent of the other riparians, enunciated by the Madrid Declaration, was both unsound and unreasonable, as it granted one riparian state veto power over the development plans of other riparian states. On the other hand, the ILA had by that time barely begun to study the issues arising out of international rivers, despite its establishment more than 70 years earlier. The two conventions adopted under the auspices of the League of Nations have little practical relevance.

As a result, the Bank had to struggle with the issues arising from projects on international waterways, and gradually, and through extensive internal deliberations and the limited implementation experience, develop its own policy for such projects. As it evolved over the last 50 years, the policy has benefited from the emerging principles of international water law, and has in turn contributed to the evolution of such principles, as we shall see in the course of the next Chapters of this Book.

---

which a reference has already been made (*supra* n. 54). The other case is the Meuse River, which was decided by the PCIJ in 1937. The Meuse River originates in France before entering Belgium, becomes a border river between Belgium and Holland and then flows into Holland before joining the Rhine. The dispute involved the digging of a canal by Belgium to divert part of the flow of the river. Because the questions at issue between the two countries were governed by the 1863 Treaty, the Court concentrated on interpretation of the Treaty, and, unfortunately, did not address the general rules of international water law. As such, the decision of the Court has not contributed much to the evolution of the principles of international water law, and the Court had actually missed an opportunity to do so. For the judgment of the Court, *see Diversion of Water from the Meuse* (Netherlands v. Belgium), Judgment No. 25 (PCIJ, Ser. A./B., No. 70, 1925) at 6.

CHAPTER 2

The Search for Policies for Projects on International Waterways

2.1 The Bank and the Competing Interests of Riparian States

As indicated in the previous Chapter, the issue of international rivers started surfacing in the late 1940s and early 1950s in a number of projects that the Bank financed or considered financing. With the emphasis of the Bank on hydropower and irrigation projects, it was inevitable that a number of those projects would involve shared rivers. Thus, the Bank started gradually to realize that the financing of projects on international rivers was delicate and complex, and demanded thorough thought and reflection. The proposed Youssef Pasha and the Ghab projects in Syria, the Aswan High Dam Project in Egypt, the Bhakra-Nagal Dam Project in India, and the Lower Sind Barrage Project in Pakistan all involved international rivers. None of those rivers—the Orontes, Euphrates, Nile, or the Indus, had, at that time, an agreement governing the uses and sharing of its waters. Actual disputes between some of the riparians of those rivers had already emerged by that time, and potential disputes with other riparians were expected to arise as a result of the proposed projects. The Bank's pipeline included other projects on international rivers such as the Kariba Hydropower Project on the Zambezi River in Rhodesia, the Lempa River Power Project in El Salvador, and the Roseiris Dam Project on the Blue Nile River in the Sudan.

The Bank was fully aware of its unique character as an international financial cooperative institution, owned and operated by its member countries. The rights and interests of each of these member countries have to be duly taken into account in Bank operations. Indeed, the Articles of Agreement of the Bank themselves clearly spell out such an obligation. Article III, Section 4(v) of the Articles of Agreement requires the Bank to " . . . act prudently in the interests both of the particular member in whose territories the project is located and of the members as a whole."[84] The

---

[84] Article III, Section 4(v) reads in full "In making or guaranteeing a loan, the Bank shall pay due regard to the prospects that the borrower, and, if the borrower is not a member, that the guarantor, will be in position to meet its obligations under the loan; and the Bank shall act prudently in the interests both of the particular member in whose territories the project is located and of the members as a whole."

interests referred to in this Article include, by necessity, the interests of all the riparian states in the shared rivers. This positive requirement on the Bank to act prudently in the interests of all its members goes beyond the general principle of international law elaborated by the IIL in the Madrid Declaration.[85] That Declaration, parts of which were confirmed by the *Trail Smelter* case,[86] prohibits acts by one riparian state that would have adverse effects on another riparian state. Indeed, based on the requirements under Article III, Section 4(v) cited above, and the nature of the Bank as an international financial cooperative institution, the Bank would not support any project that would cause adverse effects on any other riparian without such riparian's consent.

One early approach of the Bank to projects on international rivers was to distinguish between rivers over which there were disputes, and those over which no apparent disputes existed at the time the project was being considered for financing. This approach was applied as early as 1949–1950 when the Bank was considering financing the Bakhra-Nangal Project in India, and the Lower Sind Barrage Project in Pakistan. Both projects were on the Indus River System, over which a dispute had already erupted in 1948, following the partitioning of the Sub-continent.[87] Because of the dispute, and the objection of each of India and Pakistan to the Bank financing of projects on the disputed Indus River System in the territory of the other country, the Bank was not able to extend funding to either project.[88]

Similarly, the circumstances that surrounded the efforts to negotiate financing for the Aswan High Dam Project in Egypt in the early to mid-1950s were quite complex. The dispute over the sharing of the waters of the Nile River between Egypt and Sudan at that time was only one of many issues that dominated the discussion between the Bank and Egypt.[89] The Bank made a determination from the start that financing the Aswan High Dam scheme " . . . would depend on Egypt's reaching an agreement with the Sudan, and perhaps other

---

[85] *See supra* n. 64.

[86] *See supra* n. 79.

[87] For more discussion of the Indus dispute, *see infra* Chapter 4, Part 4.2 of this Book.

[88] See Mason & Asher, *supra* n. 2, at 612. *See* also Undala Z. Alam, *Water Rationality: Mediating the Indus Waters Treaty*, unpublished Ph.D. Thesis (University of Durham, United Kingdom 1998) at 97. It is worth adding that the failure of the Bank to finance either of these projects was one reason for the Bank to offer its good offices to the two countries in an attempt to find a solution to the dispute, as will be discussed in Chapter 4 of this Book.

[89] For discussion of the Aswan High Dam Project and the Bank involvement, *see* Mason & Asher, *supra* n. 2, at 627–643.

upstream countries also, about sharing the waters of the Nile."[90] However, as the discussions and studies of the project continued, the emphasis of the Bank on the need to reach an agreement became limited to Egypt and Sudan only, and no mention was made of other riparians.[91] Although the developments that took place in 1956 prevented the Bank from eventually financing the Aswan High Dam, the failure, until that time, to reach an agreement with the Sudan on the sharing of the waters of the Nile was indeed one major factor for the decision of the Bank not to finance the project.[92]

Accordingly, the Bank had to balance the competing demands and interests of the different riparians over the limited amount of the waters of the shared waterway on the one hand, and its character and responsibilities, as an international financial cooperative institution, on the other. This task has proven through years to be quite intricate and complex, requiring a very delicate and careful handling, as will be discussed in the next Chapters of this Book.

2.2   Early Approaches

The decision of the Bank not to finance projects on rivers where a dispute existed deprived the countries concerned and the Bank from pursuing development opportunities in those countries, particularly in the power and irrigation sectors. The Bank, however, had no choice. As an international financial cooperative institution, one of its concerns is dispute avoidance. Although not all international rivers were characterized by actual or apparent disputes, few rivers, if any, were governed by agreements regulating their sharing and management. Projects on those rivers still carry the risk of generating concerns, and perhaps even disputes, among the riparians, and also between some of those riparians and the Bank by virtue of the Bank funding for those projects.

---

[90] *See id.*, at 629. The authors stated that the Bank indicated to the Egyptians at one point that it could not make any final decision on the Aswan High Dam Scheme until a comprehensive study was made of the Nile development as a whole.

[91] *See id.*, at 635. This emphasis on agreement only with the Sudan could be attributed to two reasons. One of the reasons was the dispute between the two countries regarding an increase in the amount of water allocated to the Sudan. The other reason was the fact that Sudanese territory would be flooded by the Aswan High Dam, and that would require the resettlement of a large number people. *See id.*, at 632.

[92] For a discussion of the political developments, including the Egyptian-Czechoslovakian arms deal, Egypt's reorientation toward the Eastern bloc, as well as the nationalization of the Suez Canal and the 1956 war, and the effects of all those factors on the financing of the Aswan Dam, *see* Yoram Meital, *The Aswan High Dam and Revolutionary Symbolism in Egypt*, in *The Nile—Histories, Cultures, Myths*, at 219 (Haggai Erlich & Israel Gershoni, eds., Lynne Rienner Publishers 2000).

This situation was true with regard to a number of projects that the Bank was considering, including the two projects that it was studying at that time in Syria: the Youssef Pasha Project,[93] and the Ghab Project.[94] There were no actual disputes at that time on the Orontes or the Euphrates rivers. However, given the absence of an agreement on both of those rivers, the history of the relationship between the riparian states on those rivers, and the nature of the projects which involved, *inter alia*, construction of dams, there were concerns that the projects might generate protests from the other riparians. There was also concern that those projects could also ignite disputes among the riparian states, and with the Bank itself. Dispute avoidance with and among riparians as a result of a Bank-financed project was, and continues to be, a major objective for the Bank.

As mentioned earlier, the Bank started discussions with Syria in 1950 on some projects, including the Youssef Pasha Project, and the Ghab Project.[95] Those two projects were processed concurrently for about six years, ending as we shall see, in 1956. Although those two projects presented major challenges to the Bank, they also provided the Bank with an opportunity to deliberate on possible approaches for projects on international rivers. Those deliberations involved both the projects staff and the legal staff of the Bank, as well as senior management and the Executive Directors, as we shall see later.

The Youssef Pasha Project involved the construction of a multipurpose dam on the Euphrates River for (i) regulation of the river flow, and storage of part of its floods, in order to secure necessary water supply for irrigation during the dry season; and (ii) generation of hydroelectric power to be used for pump irrigation in the Euphrates Valley, water supply to the city of Aleppo and industrial development. After studies on the feasibility of the project, Syria renewed its request to the Bank for financing the multipurpose dam in 1952.

The Bank took note of the international character of the Euphrates River,[96] and the then stalled discussions on how to share it. It concluded that there was a need for some form of international cooperation between the three riparian

---

[93] Syria, Youssef Pasha Multipurpose Dam Project (1953).

[94] Syria, Draining of the Ghab Project (1953).

[95] *See supra* n. 43.

[96] The Euphrates River rises in the highlands of Turkey. It is formed by the junction of the Kara River and the Murat River, from where it flows for about 450 km to the south before it enters Syria. From there, the Euphrates traverses Syria in a southeastern direction for about 680 km, thereafter entering Iraq. Within Iraq, the Euphrates continues its southeastern direction for another 1,000 km, before it joins the Tigris northwest of Basra to flow into the Gulf at Shatt el Arab. Immediately before its estuary, the Shatt el Arab forms the frontier between Iraq and Iran. *See* Kibaroğlu *supra* n. 39.

countries (Turkey, Syria and Iraq), before financing of the project could be considered.[97] Various possible approaches were contemplated within the Bank. One approach was the conclusion of an agreement by the three riparians to establish a Euphrates Commission whose function would be to study the overall problems affecting water control and the best utilization of the waters of the Euphrates River. A similar commission, it was thought, might function in connection with the utilization of the Tigris River. When the Bank realized that the political situation was not conducive to the establishment of such a commission, other approaches were considered. One such approach was for Syria to discuss the proposed project with the Government of Iraq and, if necessary, reach some understanding on the minimum flow to be available in the Euphrates River for Iraq after the completion of the project. Similarly, the Bank concluded that this approach was not politically feasible. While the processing of the project continued, the issue of how to deal with the other riparians of the Euphrates River was overtaken in 1953 by the overall discussion of, and deliberations on similar issues related to the Ghab Project.

The Ghab Project involved the reclamation and development of the Ghab Valley in northwest Syria, through which the Orontes River flows. The River is shared by Lebanon, Syria, and Turkey.[98] The project consisted of the drainage of extensive swamp lands and their upstream protection by a retention dam and other flood control and irrigation works which included a storage dam and a

---

[97] There has been no agreement between the three riparians on the sharing of the Tigris and Euphrates. A few agreements were entered into after the collapse of the Ottoman Empire when the Tigris and Euphrates became international rivers. The first of those agreements was the Convention of December 23, 1920, between Britain and France as the Mandatory powers for Iraq and Syria. *See* the Franco-British Convention on Certain Points Connected with the Mandates for Syria and the Lebanon, Palestine and Mesopotamia, 16(3) Am. J. of Int'l L. (1922), 122–126. The Convention stated that any plans for irrigation in Syria that might diminish the flow of the Euphrates should be referred to a commission nominated by the two countries. Iraq inherited Britain's rights and obligations, including the 1920 Convention, by signing the Protocol of October 10, 1932. Syria, on the other hand, took no such step. The 1923 Treaty of Lausanne committed Turkey to consult with Iraq before carrying out any works on the Tigris or Euphrates. In 1946 the Treaty of Friendship and Neighbourly Relations between Iraq and Turkey was concluded in Ankara. Turkey agreed under the Treaty to inform Iraq of any plans for hydraulic works on the Tigris or the Euphrates (*see* 37 U.N.T.S. 281 (1949)). On July 7, 1987, Syria and Turkey concluded in Damascus the Protocol on Matters Relating to Economic Cooperation between the Syrian Arab Republic and the Republic of Turkey, *see* 1724 U.N.T.S 4 (1993). The Protocol dealt, *inter alia*, with temporary allocation of the flow of the Euphrates.

[98] The Orontes River rises in Lebanon, flows north through Syria for most of its course, forms the border between Syria and Turkey for approximately 25 miles, and then swings west and southwest through Turkey before it empties into the Mediterranean Sea. *See* Shapland, *supra* n. 47.

system of distribution canals. Altogether, the Ghab Project provided for the conversion of approximately 41,000 hectares into irrigated land.

The international aspects of the Ghab Project triggered extensive discussion within the Bank, and set in motion a search for a general approach towards projects on international waters. Because of the issue of international waters, this project was the first project to reach and be discussed by the Loan Committee.[99] The main reason for that related to the concerns raised about the likely effects of the project on Turkey, the downstream country. The Bank asked Syria for as much accurate data as possible concerning the flow of the Orontes River into Turkey, both before the project, and after the project would be completed. The Bank was keenly aware of its peculiar responsibilities as an international financial cooperative institution. Accordingly, it was alert to the need for extreme care in extending a loan to one country, for a project that might adversely affect the physical conditions in another country, without the consent of the latter, expressed in a form satisfactory to the Bank.

During the deliberations on those two projects, a number of possible generic approaches, applicable to other projects as well, were raised and discussed.[100] One of those approaches was for the Bank to assist in regional planning through the establishment of an authority to study and determine the most effective ways of utilizing the waters of the shared river among the riparian countries, and to supervise the approved uses of such a river. This approach was based largely on the principle of community of interests, discussed earlier, where the river is dealt with as an economic unit and the rights over the waters of the entire river are vested in the collective body of the riparian states.[101] However, the Bank realized, judging from the Indus and Nile negotiations underway at the time, that an agreement would take quite a while to reach, if indeed such an agreement was ever to materialize.

Another approach considered was for the Bank to require the consent of all the riparian states for the proposed project. Concerns were raised, however, that some riparians with no real interest in the project could cause delays in the starting of the project by failing to assent; or could use the request for assent to insist on quite unrelated concessions. The notion of giving veto power to one or more of the riparians was soon discarded, based on the criticism of this approach, which was proposed in 1911 by the Madrid Declaration.

---

[99] This Committee, which was established in 1953, was the highest body in the Bank dealing with operational matters; *see infra* n. 108. As will be discussed later, *see infra* n. 486, it has been renamed the Operations Committee.

[100] For a more detailed discussion of those options, *see* Raj Krishna, *International Waters in Bank Projects—History and Case Studies*, Legal Department Working Paper No. 2, at 14–25 (The World Bank 1973).

[101] *See supra*, Chapter 1, Part 1.3 of this Book.

A third approach considered was the recognition of the established uses of the shared river and protection of such uses against upstream withdrawals and downstream establishment of prior rights. This is an interesting approach as it attempted to throw into the discussion a number of concepts, including established rights, as well as potential foreclosure of future uses, which will be discussed later.[102] However, the practical difficulties of implementing this approach were recognized. Data and information provided by one riparian could be challenged by another riparian, and that could place the Bank in the difficult position of making a decision one way or another. In this way the Bank would end up being an arbiter in the case.

The fourth alternative discussed was that the facts and circumstances of each case would determine how the Bank would discharge its duty. In other words, the Bank would follow an *ad hoc* approach. This flexible approach was based on the Bank ascertaining all pertinent facts related to the project, to the best of its ability, prior to making a final determination. The approach would include requiring the prospective borrower to satisfy the Bank that the project would not result in a substantive protest from another riparian whose interests might be affected by the project. It could be that the project would have no effects whatsoever on any other riparian. Conversely, it could have effects on the other riparians and the Bank would have to seek a no-objection from those countries. In this connection the Bank would have to keep in mind frivolous or unfounded objections, or objections used for bargaining purposes.[103]

A similar approach of seeking the views of the other riparians for the Youssef Pasha Project was being considered in 1955, when it was proposed that:

> With regard to the project's international aspect, the Bank, before undertaking serious consideration of the project for financing, would have to be assured that other riparian states had been given reasonable opportunity to consider the effects of the project on their own interests and to express their views thereon, and for that purpose had been given an adequate description of the project with the necessary supporting technical data. The Bank would have to reserve its right to determine its ultimate position in the light of all the facts concerning the international aspects of the project including any views so expressed.[104]

According to this approach the other riparians should be given a reasonable opportunity to consider the effects of the projects on their interests and to express

---

[102] *See infra*, Chapter 5, Part 5.3 of this Book.

[103] *See id.*

[104] Syria, Youssef Pasha Multipurpose Dam Project, Memorandum from the Department of Operations—Asia and Middle East, SLC/0/847, at 2 (April 23, 1956).

an opinion. This meant that the other riparians would be notified of the project and the project details, and be given an opportunity to convey their comments and views. However, the opinion of the other riparians would not prevent the Bank from making its own decision on the project, taking into account all the pertinent facts. This meant that the consent of the other riparians was not essential, and that the Bank could still finance the project notwithstanding an objection from one or more of the riparians.

During those deliberations it was also observed that the objection of another riparian country could be frivolous; or based only on a remote likelihood that this country would require the water itself within a reasonable period; or designed as leverage against the other country in some unrelated matter, or intended to gain some undue advantage in the immediate issue.

With regard to this observation, concerns were raised that by making a judgment on the objection of another riparian, the Bank would be playing the role of an arbiter between two contestants. That situation, it was feared, could bring reproach to the Bank for ignoring the other riparian's interests. One way of dealing with this situation was to confirm the earlier Bank approach of not financing projects when a dispute existed between the prospective borrower and one of the riparians.

Although the deliberations raised a number of possible approaches, consensus seemed to be emerging by 1956 that the facts and circumstances of each case must determine how the Bank would handle the proposed project. This approach was referred to as the "flexible approach"[105] and, as will be discussed later, was reflected in the Bank policies for projects on international waterways that were subsequently adopted.

With regard to the Ghab Project, the Bank concluded that (i) the mere absence of an agreement between Turkey and Syria on the Orontes River or the project should not prevent the Bank from participation in the financing of the project; (ii) there was no real risk that possible uses of the water for irrigation before the river enters Syria would endanger the project; (iii) the project would control winter floods of the Orontes and thus, in this respect, benefit the entire valley to the north; and (iv) there would continue to be sufficient summer flow of the water of the Orontes River below the Ghab to provide adequate water for all land presently irrigated and for all irrigable land not presently irrigated in the valley of the Orontes River between the Ghab and the mouth of the river.

A copy of the Project report was transmitted to Turkey in August 1955. Negotiations for the project were completed in early 1956, but soon afterward, and

---

[105] *See* Raj Krishna, *The Evolution and Context of the Bank Policy for Projects on International Waterways*, in Salman & Boisson de Chazournes (eds.), *supra* n. 56, at 32.

before the project was considered by the Executive Directors, the Bank was informed of Turkey's intention to object to the project. Meanwhile, and as a result of the deliberations and developments on the processing of the Ghab Project, including the informal objection by Turkey, the Bank issued its first policy for projects on international waterways in March 1956, as discussed below. In May of that year, Turkey officially objected to the Ghab Project claiming that the Project would deprive it of the waters of the Orontes River during the irrigation season, and asked that the Project design be revised by experts from both governments. This objection was a significant milestone in the Bank's operational work on international rivers, and was indeed a major factor in shaping its rethinking of the ways and means for dealing with such projects.

As a result of Turkey's objection and demand that the Project design be revised by Syrian and Turkish experts, the Syrian government informed the Bank of its decision to postpone consideration of the Ghab Project.[106]

Thus, the processing of the Ghab Project, as well as the Youssef Pasha Project, ended before Board consideration. Nonetheless, those projects provided the Bank the opportunity to hold extensive internal deliberations on how to deal with projects on international rivers. Fueled by Turkey's objection, those deliberations resulted in paving the way for the issuance of the first Bank policy for dealing with such projects.

2.3   Birth of the First Bank Policy (1956)

As stated before, the projects on international rivers that the Bank started processing, particularly the Ghab Project in Syria, presented the Bank with both challenges and opportunities. The challenges could be summed up as how to operationalize the requirement under its Articles of Agreement that the Bank act prudently in the interests both of the particular member in whose territories the project would be located, and of the other riparians. The opportunities provided by these projects were the extensive deliberations, and the emergence of some possible approaches for dealing with those projects. Those deliberations

---

[106] No reasons for the Syrian request for the postponement of the processing of the project were included in any of the Bank records. In fact, processing of the project stopped at that stage and was not revived after that. Greg Shapland gave the following explanation: "In response to Turkish claims to the region, the Hatay province into which the Orontes flows was detached from Syria in the late 1930s when Syria was under the French mandate—essentially as a sop to Turkey to prevent it from siding with Nazi Germany. Although the Syrians must be well aware that there is no prospect of getting the region back in the foreseeable future, they maintain officially that it is part of their country. They therefore refuse to discuss the Orontes with Turkey since this could be construed as tacit recognition of Turkish sovereignty over the Hatay." *See* Shapland, *supra* n. 47, at 146.

culminated in the issuance by the Bank of the first policy dealing with those types of projects.

The policy was issued on March 6, 1956, as Operational Memorandum No. 8, and was entitled "Projects on International Inland Waterways."[107] The fact that this Operational Memorandum (OM) was number 8, which meant that it was preceded by a few other operational memoranda, and that it was issued less than 10 years after the Bank started its lending operations, underscored the urgency of the issues surrounding projects on international waterways. Indeed this was the first Bank Operational Memorandum dealing with a substantive operational matter.[108] As will be discussed later, the first environmental policy was issued by the Bank in 1984, almost 30 years after OM 8 was issued.[109]

The OM emphasized the fact that projects on international inland waterways are likely to give rise to problems which go far beyond the usual problems of project analysis in that they could affect relations not only between the Bank and its borrowers, but also among governments. The OM went on to state that these projects need special handling. However, the OM did not set forth any substantive rules for handling those projects. Instead, paragraph 1 of the OM stated that it was "vital that the Bank decide how each case will be handled before discussions have reached a stage at which it would be embarrassing to introduce new questions." In this way, the OM confirmed the "flexible approach" where the facts and circumstances of each proposed project would determine how the Bank would handle that project. It also provided the basis for dispute avoidance, which has become one major aspect of the Bank policies, procedures and operations. The main features OM 8 were:

First, it enumerated the types of projects covered, which included hydroelectric, irrigation, flood control, navigation, drainage or similar project. The term "similar project" was meant to give flexibility to the Bank to include

---

[107] The OM was approved by the Loan Committee on February 8, 1956. A copy of the OM is attached as Appendix 1 to this Book.

[108] The first set of Operational Memoranda was issued on February 26, 1953, and consisted of 10 OMs, covering a wide array of rudimentary issues such as the Staff Loan Committee Meetings and Memoranda, and Working Party Assignments. The Operational Memoranda also dealt with some basic financial issues such as Balance of Payments Reports, Debt Studies and Currency Conversion Rates. OM 8 which was issued three years later replaced in number one of those first 10 OMs which dealt with "Loan Administration Reports."

[109] *See* Operational Manual Statement (OMS) 2.36, Environmental Aspects of Bank Work (May 1984). It should be clarified, however, that the Bank established in 1972 the Office of Environmental Affairs to vet the projects that might have negative environmental effects, and advise staff on how to deal with them. For more details on this Office, see Ibrahim F. I. Shihata, *The World Bank and the Environment: A Legal Perspective*, 16 MD. J. Int'l L & TR 1, (4) (1992).

projects not enumerated in the OM, but which could affect the international waterway in question.

Second, the OM also went beyond the term "international rivers" used by the Bank until the OM was issued. As may be noted, the international waters issues that the Bank faced until that time all related to shared rivers, and the Bank had not been faced by then with issues of transboundary lakes. Yet, OM 8 used the term "international inland waterways" as the title of the OM.

Third, it is worth noting that the OM chose the term "waterways" over the term "watercourses" used by the IIL in the Madrid Declaration. It may be recalled that the first international instruments to use the term "waterways" was the Barcelona Convention which is entitled "Convention and Statute on the Regime of Navigable Waterways of International Concern."[110] As will be noted later, this term continued to be used in all the successive policies issued by the Bank thereafter.

Fourth, the OM defined international inland waterways to include a river, canal, lake or other inland waterway "which forms a boundary between or flows through two or more countries." Accordingly, the OM made it clear that the definition included successive rivers, as well as boundary rivers and lakes.

Fifth, the countries sharing the waterways did not need to be members of the Bank. Paragraph 2(a) of the OM extended its application to all riparians of the shared inland waterway, "whether members or non-members of the Bank." This is an interesting, albeit necessary, feature of the OM. It went beyond the responsibilities of the Bank to all its members under the Articles of Agreement, and emphasized application of the OM to non-member riparian states as well. It is necessary because of the general obligation under international law to avoid harm to all other riparians. This is a basic obligation that the Bank could not possibly overlook. By making an explicit reference to non-members, the Bank is simply emphasizing and underscoring this obligation. This emphasis on the application of the policy to non-members of the Bank would remain as one of the main features in all subsequent Bank directives for projects on international waterways. It is worth adding that the Bank membership had grown from 38 countries in 1946, to 58 in 1956.[111]

---

[110] *See supra* n. 55. Article 1 of the Statute on the Regime of Navigable Waterways defines navigable waterways of international concern to include "All parts which are naturally navigable to and from the sea of a waterway which in its course, naturally navigable to and from the sea, separates or traverses different States, and also any part of any other waterway, naturally navigable to and from the sea, which connects with the sea a waterway naturally navigable which separates or traverses different States."

[111] *See supra* n. 4. For the membership of the Bank in 1956, *see* International Bank for Reconstruction and Development, *Eleventh Annual Report*, 1955–56, at 27.

Sixth, the OM did not set forth any specific policies or substantive rules and procedures to be followed for these projects with regard to the prospective borrower and the other riparians. Rather, it stated that if the project was covered by the OM, then:

(b) . . . the management should be informed promptly and the Working Party for the project and the interested department should make it their first order of business to propose and obtain management approval of a procedure for dealing with the international aspects of the project.

(c) No steps should be taken to investigate the merits of the project or to process the project without prior approval from the management. This requirement is to be observed even if the international aspects appear to be covered by international agreement or if there appears to be *prima facie* evidence that the project cannot be adversely affected by, or will not adversely affect, up-stream or down-stream riparian states.[112]

The fact that the OM did not include any specific policies or procedures, but left it to the Working Party to propose and obtain management approval for such procedures indicated that the Bank had decided to follow the "flexible approach" discussed above, and to deal with each project based on its facts and circumstances. It is worth noting that the requirement of proposing and obtaining management approval of a procedure for dealing with the project applied to cases where the issues were covered by an international agreement, as well as to cases where there might be *prima facie* evidence that the project could not be adversely affected by, or would not adversely affect, up-stream or downstream riparian states. In this way, the OM left no exceptions for cases where the Working Party could proceed with the project before informing Bank management.

Seventh, as indicated above, the OM addressed both situations: the project not being adversely affected by, or adversely affecting, upstream or downstream riparian states.[113] This statement raises two issues.

The first is whether the Bank was concerned only with adverse effects on downstream riparians by upstream riparians, or whether it contemplated that downstream riparians could also affect upstream riparians? It is interesting that the deliberations within the Bank prior to the issuance of the OM raised the possibility that Bank-financed projects could assist borrowers in establishing water rights that could be later claimed as established or acquired rights. The establishment of water rights and claiming them later as acquired rights basically relates to the potential foreclosure of future uses of such

---

[112] *See* paragraph 2(b) and 2(c) of OM 8.

[113] *See* paragraph 2(c) of OM 8.

waters, as will be discussed later.[114] This concept fosters the argument that downstream riparians can adversely affect upstream riparians by foreclosing the future uses of the waters of the shared river. It may not be easy to argue that this concept was really comprehended at that time by Bank staff and management, and that it was indeed intended by the above paragraph in the OM. However, the deliberations as well as the content of the OM indicate that the issue was being thought about.

The second issue related to the notion that a project may be adversely affected by the other riparians. The most likely possibility in this connection is that future projects in one upstream riparian may affect Bank-financed projects in a downstream riparian, such as by decreasing the water flow to this downstream riparian. This situation can only be dealt with by trying to ascertain the intentions and plans of those upstream riparians. Ascertaining such intentions is not a simple task, and it may not be easy to undertake given that such riparians would be under no obligation to disclose their future plans, even if they had prepared such plans.

Regardless of the intentions of the drafters of the OM on both issues, one conclusion that can be drawn from the language of the OM is that the issues concerning projects on international waterways were at that time, and continue to be, intricate and complex. More than 50 years after the issuance of the first Bank policy on the topic, a number of controversies and ambiguities still surround the basic principles of international water law, as well as the Bank policy.

As discussed earlier, by the time it issued OM 8, the Bank had deliberated a number of approaches for dealing with projects on international waterways. The first such approach was that the Bank would not finance a project where a dispute over the shared river clearly existed.[115] The second was to require an agreement between the concerned riparians, including setting up an authority or commission for the river. The third was to require consent of all the riparians. The fourth approach, which gradually gained ground, was that the facts and circumstances of each case would determine how the Bank would discharge its duty to the prospective borrower and the other riparians. Those facts and circumstances included, *inter alia*, the existence or conclusion of an agreement on the uses of the waters of the river; the interests of downstream riparians were not being adversely affected by the project; the downstream riparians had no objection to the project; or that such

---

[114] *See infra*, Chapter 5, Part 5.3 of this Book.

[115] As indicated before, this was evidenced by the refusal of the Bank to finance a number of projects, including the Aswan High Dam Project in Egypt, the Bhakra-Nagal Dam Project in Pakistan, and the Lower Sind Barrage Project in India. *See supra*, Chapter 1, Part 1.1 of this Book.

objection was without merit. Obtaining a response from a riparian, whether such a response is consent or an objection, would only take place if such a riparian was notified of the project as had happened with the Ghab Project. As discussed earlier, in that case the Bank notified Turkey, and provided the technical data and information about the project.

Thus, the OM and the flexible approach it adopted set in motion a process of notification of the other riparians of Bank-financed projects on international waterways. This process has resulted in extensive and rich legal literature in the field, as will be discussed later. This literature has been influenced by the existing and emerging principles of international water law, and in turn contributed to the development of those principles.

OM 8 was, without doubt, a major milestone in the operational work of the Bank in the field of shared waterways. It was issued despite the paucity of international law with regards to international rivers, and the lack of guiding international rules and principles in the mid-1950s. As discussed earlier, even the work of the IIL and the ILA on shared watercourses was at that time in its infancy. It is true that the OM did not include any substantive principles and policies for dealing with the intricate issues posed by projects on international waterways. However, the OM underscored the obligation of the Bank of avoiding disputes and seeking an approach for dealing with the international aspects of the project that would take into account the interests of all the riparians. The Bank itself had by that time elaborated a number of possible approaches for dealing with each project. As Raj Krishna noted, the OM "established an early warning system by requiring the Bank's management to be informed promptly of any project which involved the use of an international watercourse. No steps were to be taken to investigate the merits, or process a project without management approval of a procedure for dealing with the international aspects of the project."[116] The different approaches proposed and followed by the Bank for the varying projects that the Bank financed on international waterways gradually constituted the building blocks for the evolution of the successive Bank policies in this field.

2.4   The 1965 Updates to the 1956 Policy

Operational Memorandum No. 8, issued in 1956, continued in force until it was replaced by Operational Memorandum 5.05 which was issued on January 1, 1965.[117] A number of factors prompted the Bank to undertake the revisions to OM 8. Membership of the Bank had increased considerably from 58 members

---

[116] *See* Krishna, *supra* n. 105, at 32–33.

[117] Operational Memorandum 5.05 is attached as Appendix 2 to this Book.

in 1956, to 103 in 1965.[118] This followed attainment of independence by many countries, particularly in Africa, many of which joined the World Bank immediately thereafter. Most of those countries were attracted to the Bank by, *inter alia*, the economic and sector work undertaken by the Bank, as well as the interest-free credits provided by IDA. As may be recalled, IDA was established in 1960 to assist such poor member countries.[119] The establishment of IDA and the expansion in the membership of both the Bank and IDA resulted in a considerable increase in the lending program of the two institutions, and in turn, in new types of projects in their operational work, including projects on international waterways. Furthermore, the implementation experience of OM 8 during the years 1956–65 revealed areas where OM 8 needed clarifications and updates. Thus, less than 10 years after OM 8 was issued, it was replaced by OM 5.05. It should be added, however, that despite growth in the membership of the Bank, OM 5.05 continued the approach started by OM 8 in 1956; it was applicable to waterways flowing through two or more countries "whether members of the Bank/IDA or not."[120]

OMS 5.05 introduced a number of changes to OM 8. Those changes can be summarized in the following:

First, the title of the OM was changed. The previous title of OM 8 "Projects on International Inland Waterways" was replaced by the title "Projects on International Waters." This change was prompted by the inclusion of a new paragraph, discussed below, adding a new category of international waterways comprising bays, gulfs, straits and channels to the waterways covered under OM 8. As such, the 1965 OM no longer dealt only with inland waterways such as rivers, lakes and canals. Its scope was expanded to include semi-enclosed coastal waters such as bays, gulfs, straits and channels. Thus, both inland waters as well as semi-enclosed coastal waters were covered under the 1965

---

[118] For the membership of the Bank in 1956 *see supra* n. 111. For the membership in 1965, *see* International Bank for Reconstruction and Development and International Development Association, 1965–1966 *Annual Report*, at 24.

[119] As mentioned earlier, *supra* n. 1, IDA is the arm of the World Bank that helps the world's poorest countries. While IBRD raises most of its funds on the world's financial markets, IDA is funded largely by contributions from the governments of its richer member countries. Additional funds come from IBRD's income and from borrowers' repayments of earlier IDA credits. The first IDA credits were approved in 1961 to Chile, Honduras, India and Sudan. Thirty-four countries have graduated from IDA throughout its history. A few of those countries have since "reverse graduated," or reentered IDA. For more information on IDA *see*: http://web.worldbank.org/WBSITE/EXTERNAL/EXTABOUTUS/IDA/0,,menuPK:51235 940~pagePK:118644~piPK:51236156~theSitePK:73154,00.html. The term "Bank" will henceforth be used to include both IBRD and IDA.

[120] *See* paragraph 1(b)(i) of OM 5.05.

OM. This change in turn resulted in a new title to the 1965 OM to encompass both types of waterways.

Second, the 1965 OM was divided into two paragraphs. The first paragraph dealt with and was entitled "Projects on International Inland Waterways," and reiterated the entire OM 8, with the changes discussed below. The second paragraph dealt with, and was entitled "Projects Involving Other International Waters."

Third, the list of projects enumerated in the 1965 OM included two new types of projects—sewage and industrial projects—in addition to those included in the 1956 OM.[121]

Fourth, the effects of the proposed projects extended from use of the waters, and included pollution of those waters. As such, both water quantity as well as quality issues were included in the 1965 OM.

Fifth, a new paragraph 2 was added. This paragraph stated that the procedures set forth in the preceding paragraph should also be followed in the case of any project involving such international waters as bays, gulfs, straits, or channels bound by several states or, if within one state, recognized as necessary channels of communication between the open sea and other states.

This paragraph basically added semi-enclosed coastal waters to the policy. The addition of such waters to inland waterways may seem confusing. The 1956 OM was meant to address the pressing and complex issues of projects on shared rivers and lakes. A number of those rivers and lakes have been characterized by the competing demands of the different riparian states. The Bank policy for projects on international inland waterways was meant to ensure that the Bank would act prudently in the interests of all the states concerned, and try to avoid generating disputes, or exacerbating existing ones. Thus, the immediate question that arises is what prompted the Bank to add such semi-enclosed coastal waters to its policy on inland waters, and what purpose was the new addition expected to serve?

The Bank had by the early 1960s started financing navigation,[122] industrial, and sewage projects. The latter two types of projects were added to the list of projects in the 1965 OM. These projects could affect the quality of the waters of bays, gulfs, straits or channels by polluting such waters. Accordingly, the concern was more on quality, than quantity. The Bank wanted to ensure that the projects it planned to finance would not adversely affect the quality of the waters of those bays, gulfs,

---

[121] The list of the projects enumerated in paragraph (b)(i) of OM 5.05 included "hydroelectric, irrigation, flood control, navigation, drainage, sewage, industrial or similar project."

[122] It should be clarified that the Bank considers navigation projects to include not only the construction of ports, but also the financing of ships.

straits and channels. It should be recalled that when the semi-enclosed coastal waters were included in the 1965 OM, the Bank still did not have a policy on the environment; the first such policy was issued in 1984.[123] Obviously these types of waters, and projects, are best dealt with through an environmental impact assessment which would identify any adverse impacts caused by the project, and recommend ways for mitigating such impacts, and not through the policy for projects on international waterways. Since the Bank did not have a policy on the environment in 1965, semi-enclosed coastal waters were added to the policy on inland waterways.

One other justification given for adding these types of waters was dispute avoidance among the littoral states, as well as between the Bank and those states, on the issue of the pollution of these waters through Bank-financed projects. Disputes have arisen among a number of countries on the boundaries of their semi-enclosed coastal waters, and financing projects that might affect them was bound to exacerbate these disputes.[124] As we will see later, those types of semi-enclosed coastal waters continue to be part of the Bank policy for projects on international waterways even after the Bank adopted its policies on environmental assessment. Moreover, the scope of these types of waters has been expanded, as we shall see in the next Part of this Chapter.

In addition to the above, the 1965 OM included a new paragraph concerning the presentation of loans and/or credits to the Executive Directors of the Bank and IDA. The paragraph stated that "when presenting loan or credit projects on international waters to the Executive Directors, both the Appraisal Report and the Report and Recommendation of the President should state that the Bank/IDA has considered the international aspects of the project and is satisfied that:

(i)   the issues involved are covered by appropriate arrangements between the borrower and other riparians; or

(ii) the other riparians have stated (to the borrower or to the Bank/IDA) that they have no objection to the project; or

(iii) the project is not harmful to the interests of other riparians and their absence of express consent is immaterial or their objections are not justified."[125]

---

[123] *See* OMS 2.36, *supra* n. 109.

[124] The Gulf of Fonseca is one such example. The Gulf is bounded by El Salvador, Honduras and Nicaragua. Honduras and Nicaragua had for a long time disputed their maritime boundaries, and the issue had to be settled by the ICJ in 2007. *See Case Concerning Territorial and Maritime Dispute between Nicaragua and Honduras in the Caribbean Sea* (Nicaragua v. Honduras), available at: http://www.icj-cij.org/docket/index.php?p1=3&p2=3&code=nh&case=120&k=14.

[125] *See* paragraph 3 of OM 5.05.

The paragraph added substantive elements to the basic procedural rules included in OM 8. It basically codified the approaches that were deliberated by Bank staff and management in the 1950s during the processing of the Ghab and Youssef Pasha projects. Henceforth every Bank-financed project on international waters would fall under one of the above three categories. As a first step, the Bank would need to determine if the shared waterway or the project was covered by an agreement between the borrower and the other riparians. If the project was not covered by such agreement, then those riparians would have to convey to the borrower or the Bank their no-objection to the project. Although the OM did not require notification explicitly, there was no way for the other riparians to give a no-objection to the project unless they were already notified.

The third step proposed by the 1965 OM would follow as a result of the notification, and that either no express consent would be received, or an objection was conveyed to the Bank or the borrower. If the Bank had determined that the project was not harmful to the interests of the riparians, then the absence of express consent would be immaterial. For the same reason, the objection would not be justified. In this way the Bank became the final arbiter of the differences in opinion with the other riparians on the effects of the project on the shared waterway.

That step codified the Bank's earlier approach of not financing projects that could cause adverse effects to the other riparians. The 1965 OM used the phrase "harmful to the interests of other riparians" which is not different in substance from "will not adversely affect" also used in the OM itself.[126] Henceforth, the Bank approach to projects on international waterways would center on the obligation not to cause harm, or the principle of *sic utere tuo ut alienum non laedas*.[127]

Since there were very few cases of river basin agreements encompassing all the riparians, notification of other riparians of the proposed project started to become the basic requirement of Bank-financed projects on international waterways. No doubt, this was a major milestone in the evolution of the Bank policy, and a significant step in clarifying the hitherto uncodified Bank policy in this field.

It should be added in this connection that the IIL issued in 1961 its famous Salzburg Resolution, which included articles on notification.[128] It is most likely

---

[126] *See* paragraph 1(b)(ii), and paragraph 3 of OM 5.05.

[127] *See supra* n. 67.

[128] The IIL 1961 Salzburg Resolution is entitled "Utilization of Non-Maritime International Waters (Except for Navigation), *see supra* n. 69. Article 4 of the Resolution states that "No State can undertake works or utilization of the waters of a watercourse or hydrographic basin which seriously affect the possibility of utilization of the same waters by other States except on condition of assuring them the enjoyment of the advantages to which they are entitled under Article 3, as well as adequate compensation for any loss or damage." Article 5 states that "Works or utilization referred to in the preceding article may

that the issuance of that resolution had influenced the changes and clarifications to the Bank policy as enunciated in OM 5.05 of 1965. This Resolution will be discussed in more details in the next Chapter.

The new paragraph of OM 5.05, however, raises a number of questions. The first question concerns the entity that would send the notification to the other riparians. The OM did not provide an explicit answer. It stated that the other riparians had stated their no objection to the project to the borrower or to the Bank. This would mean that the notification could have been provided by either the borrower or the Bank. It may be recalled that Turkey was notified of the Ghab Project by the Bank, and not Syria. It should also be added that the notification was communicated through the Executive Director representing Turkey, and not directly to the Turkish government.

The second question relates to projects that were not expected to be harmful to the interests of the other riparians. Should those riparians still be notified of those projects? Sub-paragraph (iii) of the OM stated that "the project is not harmful to the interests of other riparians and their absence of express consent is immaterial or their objections are not justified." It seems that the OM was meant, and should be read, to require notification for all types of projects including those that were not harmful to the interests of other riparians, and that in such cases the absence of consent would be immaterial, and the objection unjustified. However, as will be discussed in the next Chapter, this sub-paragraph was interpreted in some instances as not requiring notification when the project was not harmful to the interest of other riparians.

The third question relates to how the Bank would handle an objection from one or more riparians. The OM was silent on this issue. It seemed that the Bank intentionally left this issue unaddressed, and would deal with such objections on a case-by-case basis, depending on the nature of the project and that of the objection.[129]

OM 5.05 was reissued in March 1971 as OM 2.22, with no changes to the content. OM 2.22 was again reissued as Operational Manual Statement (OMS) 2.32 in October 1977, with only some nomenclature changes to reflect regional

---

not be undertaken except after previous notice to interested States." Article 5 of the Salzburg Resolution clarifies the steps to be taken when one of the interested states objects to such works or utilizations. For the full text of the Resolution, *see* 49 Annuaire de l'Institut de Droit International 370 (1961); *see* also, 56 Am. J. Int'l L. 737 (1962).

[129] As will be discussed later, the Salzburg Resolution addresses the issue of objection by one or more of the riparian states in Article 6 which reads "In case objection is made, the States will enter into negotiations with a view to reaching an agreement within a reasonable time. For this purpose, it is desirable that the States in disagreement should have recourse to technical experts and, should occasion arise, to commissions and appropriate agencies in order to arrive at solutions assuring the greatest advantage to all concerned."

organization. Accordingly, the procedural and substantive rules set forth in 1965 in OM 5.05 would continue to govern the financing of projects on international waters until those rules were significantly revised in 1985, as we shall see in the next Chapter.

OM 8 issued in 1956 brought the subject of projects on international waterways to the forefront of the Bank operational work, and established an early warning system. OM 5.05 of 1965 added some needed substantive and procedural provisions, and expanded the scope of the international waterways and projects covered thereunder. Both memoranda are indeed milestones in the process of evolution of the Bank policy for projects on international waters.

CHAPTER **3**

The Road to the Current Bank Policy

3.1 Implementation Experience until 1985

The previous Chapters discussed and analyzed the tremendous efforts that the Bank exerted in its early years to address the difficulties encountered with projects on international waterways. Those difficulties were enormous. Raj Krishna listed a number of them, including "the denial of the existence of a riparian issue; the variety of relationships between watercourse states; a reluctance to notify other watercourse states; the difficulty in concluding appropriate arrangements; the refusal to acknowledge potential harm to other watercourse states; the diversity of watercourses; the existence of disputes; the involvement of entities with unsettled status or non-members."[130]

The deliberations within the Bank during its early years on how to deal with the international water aspects of these projects identified a number of possible approaches. Those deliberations kept in mind the character of the Bank as an international financial cooperative institution, and the requirement under its Articles of Agreement to act prudently in the interests of all its members. They also took account of the emerging principles of international law in this field. Accordingly, a general rule was laid out that the Bank would not finance a project that would cause appreciable harm to another riparian.

One approach that was deliberated required that an agreement be concluded by the riparians before the Bank would finance a project on an international waterway. This agreement could be project specific, or for wider regional planning. However, the Bank noted that most of the international rivers had no agreements regulating their shared use and management. Moreover, the negotiations between India and Pakistan on the Indus River went on for about ten years,[131] and those on the Nile River between Egypt and Sudan continued for five years, before the agreements were concluded.[132] Thus, the idea of requiring an agreement, or

---

[130] *See* Raj Krishna, *International Watercourses: World Bank Experience and Policy*, in *Water in the Middle East: Legal, Political and Commercial Implications*, 35–36 (J. A. Allan & Chibli Mallat, eds., I. B. Tauris Publishers 1995).

[131] For the negotiations between India and Pakistan and the conclusion of the Indus Waters Treaty, *see infra* Chapter 4, Part 4.2 of this Book.

[132] Negotiations between Egypt and Sudan on the division of the Nile waters, and construction of the Aswan High Dam by Egypt, and Roseiris Dam by Sudan started in 1954.

regional planning, showed its limitations, and indeed difficulties. Furthermore, most waterways are shared by more than the prospective borrower and the party that may indicate its concerns about the project. Should the agreement be between those two states only, or should it extend to all the riparians? Even, in the few cases where an agreement was eventually concluded between the disputing riparians, it took a long time and effort to do so.

The Semry Rice Project in Cameroon was a case that manifested those difficulties.[133] The project included, *inter alia*, the provision for reinforcement of a 30-mile long dike, construction of pumping stations, and construction of irrigation and drainage networks, all on the Logone River. The River forms the boundary between Cameroon and Chad, and the two parties had concluded an agreement on the sharing of the waters of the river during the dry season. However, it was realized that the length and height of the dike on the Cameroon side of the river would increase the flooding of lands in Chad unless the dikes there were also repaired. The Bank brought the matter to the attention of the two governments of Cameroon and Chad. Consequently, an agreement on these points and other matters relating to the use of the Logone River waters was concluded on August 20, 1970.[134] It was noted, however, that the agreement was valid for only 10 years, and Bank staff were concerned that this period was far shorter than the 40-year period of the credit to Cameroon.[135] Further discussion within the Bank and with Cameroon took place, and after some time the agreement was modified by exchange of letters extending its duration from 10 to 40 years.[136] This whole process took about two years and was the main reason for the

---

In 1959 the two parties signed the Agreement between the Republic of Sudan and the United Arab Republic for the Full Utilization of the Nile Waters. As a follow-up to the Agreement, the two parties signed on January 17, 1960, the "Protocol Concerning the Establishment of the Permanent Joint Technical Commission." For both instruments, *see* U.N.L.S. B/12 (International Rivers), at 143. *See* also 453 U.N.T.S. 64 (1963).

[133] *See* Cameroon, Semry Rice Project (P000320, 1972).

[134] *Protocole d'Accord Concernant les Aménagements Hydrauliques sur le Logone*, Agreement between the Federal Republic of Cameroon and the Republic of Chad (August 20, 1970) (the Logone Agreement).

[135] At that time most Credits were extended for 40 years, including a 10-year grace period.

[136] The Development Credit Agreement between IDA and Cameroon included a clause making the termination, amendment, waiver or suspension of the Logone Agreement by Cameroon, without the prior approval of the IDA, a condition of suspension. *See* Article V of the Development Credit Agreement between the Federal Republic of Cameroon and the International Development Association, Semry Rice Project, *supra* n. 133. This clause had its limitations too, because if the agreement were to be terminated, that would be done by Chad, and not by Cameroon.

delay in project processing.[137] The process showed clearly the difficulties that could be encountered when agreements were to be concluded. Henceforth, the Bank would finance projects requiring agreements with other riparians only in exceptional cases.

Another approach that was abandoned immediately after it was deliberated was requiring the consent of all the riparians for the project. It was realized that consent could be withheld for reasons that may not be related to the project. Recognition of established uses and working through them was also considered but the practical difficulties surrounding this approach soon surfaced. This led to the *ad hoc* or flexible approach where the facts and circumstances of each project and waterway would determine how the project would be handled.

With its arrival at this overall approach, the Bank issued its first OM in 1956, establishing an early warning system for projects on international waterways. In essence this system required that Bank management be informed of this type of projects, and that no steps be taken to investigate such a project prior to management approval of procedures for dealing with its international aspects. In 1965, less than 10 years after OM 8 was issued, OM 5.05 would add some substance to the procedures laid out in OM 8. That OM required investigating whether an agreement between the borrower and the other riparians existed, and if so, whether the issues involved were covered by that agreement.

Moreover, although the term "notification" was not mentioned in the 1965 OM, the OM clearly set forth the requirement of notification for projects on international waters financed by the Bank. The notified riparians would get the opportunity to express their views, but they had no veto power over the project. The Bank could still proceed with the project when such objections were not justified, or had no merit.

Those two Operational Memoranda provided major clarifications to the procedures and substance for projects on international waterways. Nevertheless, the implementation experience showed some gaps, ambiguities and cases of

---

[137] This whole process missed an important point. The Logone River is a tributary of the Shari (or Chari) River which flows into Lake Chad. Thus, the process should have involved not just Cameroon and Chad, but also the other riparians of Lake Chad, namely Niger and Nigeria. In 1970 when processing of the project started, the Lake Chad Basin Commission (LCBC) has already come into existence, having been established according to the 1964 Convention and Statutes Relating to the Development of the Chad Basin. The Convention and Statutes were signed at Fort Lamy, Chad, on May 22, 1964, by Cameroon, Chad, Niger and Nigeria. The Central African Republic joined in 1994. For a copy of the Convention and Statutes, *see* Food and Agriculture Organization (FAO), *Treaties Concerning the Non-Navigational Uses of International Watercourses—Africa*, 10, 61 FAO Legislative Series (FAO 1997).

oversight. One of the issues faced was the definition of the terms "river" and "lake." Would a project in a tributary or sub-tributary be covered by the OM? What about a tributary that runs exclusively in one country? What about national rivers that flow into a shared lake? As will be discussed later, the IIL adopted the concept of "watercourses," while the ILA dealt with the concept of "drainage basin." The Bank started with the term "waterways" in 1956, moved to the term "waters" in 1965, and then reverted again to the term "waterways," as will be discussed later.

A number of issues were faced with regard to notification. If a project on international waters would cause no adverse effects on any of the other riparians, should those riparians still be notified, or would the project be covered by the statement in the OM that "the project is not harmful to the interests of other riparians"?[138] Should downstream riparians be required to notify upstream riparians?[139] Other questions that arose included who would undertake the notification (the Bank or the borrower), as well as notification to river basin management organizations under existing agreements.[140] Questions also arose as to who would make the determination as to whether the project was covered under an existing agreement. Other issues that arose included the length of period for reply, and what should be done when no reply was received by that time, as well as how to deal with non-members—should they also be notified?

Although objections from notified riparians were rare, they did raise a host of issues when they occurred. Some of these issues related to the type of information

---

[138] In the India, Assam Agricultural Credit Project (P009713, 1977) no notification was undertaken. The reason given was that no major or minor tubewells were proposed, but only shallow tubewells and river lift pumps. The river in question was the Ganges River which is shared by China, Nepal, India and Bangladesh.

[139] The question arose in the Somalia, Trans Juba Livestock Project (P002443, 1974). The Juba River originates in Ethiopia and flows through Somalia before emptying into the Indian Ocean. The Project included the development, *inter alia*, of stock routes and water points. Ethiopia was not notified, and the reason given was that the amount of water used was minimal, and that Ethiopia is an upstream riparian. However, as we will see later (*infra*, Chapter 6, Part 6.3 of this Book), Ethiopia was notified of the Baardhere Dam on the Juba River in Somalia.

[140] This question arose in connection with Senegal, in the Senegal, River Polders Project (P002282, 1972). The river basin organization in this Project was the Organization for the Management of the Senegal River (known by its French name, *Organisation pour la mise en valeur du fleuve Sénégal*, or OMVS) established under the Convention Relative au Statut du Fleuve Sénégal (between Senegal, Mali and Mauritania, on March 11, 1972). The OMVS as well as Mali and Mauritania were notified. Guinea, which is a riparian of the Senegal River, but was not at that time a party to the Convention, was not notified. For a copy of the Convention *see* FAO Legislative Study, *supra* n. 137, at 24–31.

needed to allay the concerns of the objecting riparian,[141] the situations where the Bank could proceed with the processing of the project despite an objection,[142] and cases where a third party might be involved to give an independent opinion. A more difficult situation could arise if the notified riparian challenged the assessment of the Bank that the project would have no adverse effects on that riparian. Indeed, the revisions to OMS 2.32 of 1977 (which replaced OM 2.22 of 1971, which in turn replaced OM 5.05 of 1965) took place in 1984, as will be discussed below, as a result of such an objection and challenge to the Bank assessment that the project would have no adverse effects.

The Igdir-Aksu-Eregli-Ercis Irrigation Project (Igdir-Aksu Project) in Turkey involved the financing of four irrigation sub-projects that were expected to bring about 113,000 hectares of dry land or inadequately irrigated areas, under improved irrigation.[143] The Project included, *inter alia*, the construction of additional irrigation canals, intakes, drainage works, and rehabilitation and improvement of an existing irrigation and drainage system. One of the sub-projects, the Igdir sub-project, involved utilization of water from the Aras River. This River formed at that time the border between Turkey and the Union of Soviet Socialist Republics (USSR or Soviet Union), and further downstream, formed the border between the USSR and Iran. A Convention (with a Protocol attached thereto) concluded in 1927 between Turkey and the USSR[144] was considered by the Bank to be a satisfactory arrangement for the project, and hence the USSR was not notified of the project.[145] However, Iran was not a party to this or any other

---

[141] In the Iraq, Lower Khalis Irrigation Project (P005235, 1973), Iran was notified since the project involved the use of waters from the Diyala River. The River originates in Iran and flows through Iraq where it joins the Tigris which Iraq shares with Syria and Turkey; *see supra* n, 39. The Project also involved the use of waters from the Tigris River. Iran objected to the project following notification. The Bank entered into lengthy negotiations with Iran, and assured Iran that the project would not harm its interests. Subsequently Iran gave its no-objection to the project.

[142] In the Syria, Balikh Irrigation Project on the Euphrates River (P005546, 1974), the Bank decided to go ahead with the project despite Iraq's objection on the ground that the project would not cause adverse effects to Iraq.

[143] Turkey, Igdir-Aksu-Eregli-Ercis (IAEE) Irrigation Project (P008950, 1984).

[144] *See* Convention and Protocol between the Union of Soviet Socialist Republics and Turkey regarding the use of frontier waters, signed at Kars, January 1927; 926, 127 British and Foreign State Papers 1927 (H.M. Stationary Office 1932).

[145] There were some views within the Bank that even if the Bank had concluded that the agreement was not a satisfactory arrangement for the project, the USSR would not have been notified of the project because it was not a member of the Bank at that time. Those views were based on the premise that the Bank only needed to ensure that the project would not cause appreciable harm to such a non-member, and that there was no notification requirement.

agreement on the Aras with Turkey, although Iran had concluded an agreement in 1957 with the USSR on the Aras River.[146]

Bank staff was of the opinion that the storage of additional water from the Aras River for the Igdir sub-project would not have any significant adverse effects on Iran. These findings were conveyed to Iran, and at first no response was received from it. However, the Executive Director representing Iran requested that the consideration of the proposed project by the Executive Directors of the Bank scheduled for June 9, 1983, be postponed until June 14, 1983. Iran claimed that the project might have a detrimental effect on Iran, and asked for further information and technical data. At the meeting on June 14, 1983, the Executive Directors decided to defer consideration of the proposed loan for another month, until July 12, in order to allow the Government of Iran more time to examine the technical data on the project.

Meanwhile, the Iranian Government reiterated concerns that the project would adversely affect operation of one of its dams on the Aras River during possible drought. The Iranian Government proposed that consideration of the loan should be postponed until a trilateral commission including Iran, the USSR and Turkey had reviewed the project. While the proposal of a trilateral commission was rejected, project consideration by the Executive Directors was further delayed. Iran continued to insist that the project would cause adverse effects to its interests. Iran asked for more information which was provided, and the date of consideration of the project of October 4, 1983 was postponed further. By that time Bank staff reconfirmed that there would be no adverse effects on Iran,[147] and recommended proceeding with the processing of the proposed project under paragraph 4(c) of OMS 2.32, which stated that "the project is not harmful to the interests of other riparians and their absence of express consent is immaterial or their objections are not justified."

Correspondence between the Bank and Iran continued, and a Bank mission visited Iran in April 1984 to discuss the project and Iran's objection. The project

---

[146] *See* Agreement between Iran and the Soviet Union for the Joint Utilization of the Frontier Parts of the Aras and Atrak for Irrigation and Generation of Power, Tehran, August 1957, 428; 163 British and Foreign State Papers, 1957–58, (H.M. Stationary Office 1966).

[147] The main issue in the case of Iran was that the project would cause a slight (on average, about three percent) reduction in energy production particularly in years of low river flows. Bank staff believed that reduction to be acceptable for two reasons. First, it was only a small proportion of the energy production from the Aras Hydropower Project and Iran's economic benefits from the river (i.e., about 0.3 percent of total economic benefits). At that time, about 10 percent of Iran's economic benefits from the Aras River waters came from power production and 90 percent from irrigation. The second reason was that the reduction in energy production in Iran, due to Turkey proposing to use part of its reasonable and equitable share of the Aras River waters, was fully predictable by Iran at the time of the construction of the original Aras project, as it had been predicted for subsequent projects.

was finally presented to and approved by the Executive Directors on June 5, 1984, one year after the original date planned for its presentation. The approval was part of a compromise that was proposed by the Alternate Executive Director for Iran. The compromise consisted of three elements: (i) approval of the project as presented by Bank management; (ii) use by Bank management of its good offices to promote an agreement between Turkey and Iran over the Aras River; and (iii) preparation by Bank management of a paper for the purposes of review of the present guidelines for projects on international waters. This latter element was included because a number of Executive Directors came to recognize that those guidelines were not adequate to deal with the complex issues raised by projects of the nature of the Igdir-Aksu Project.

Accordingly, it took one full year to deal with the objection of Iran to the project, and its challenges to the Bank staff determination that the project would not cause adverse effects to its interests.[148] The project was repeatedly taken out of the agenda of the Executive Directors of the Bank, and more data and information was requested by, and supplied to Iran. Turkey assisted in supplying such data and information, but kept pressing for presentation of the project to the Executive Directors. As a result, the Bank became an arbiter in the dispute over the project and its impacts, and the process gradually turned into negotiations over larger issues, not just the project, namely the Aras River and the Bank policy for projects on international waters.

Turkey got the project approved by the Executive Directors. On the other hand, Iran got a promise that the Bank would use its good offices to promote an agreement over the Aras River between the Iran and Turkey. Furthermore, Iran managed also to obtain a recognition that the Bank guidelines for projects on international waters were not adequate to deal with the complex issues raised by projects of that nature, and that Bank management would prepare a paper for the purpose of the review of those guidelines.

The Bank did take initial steps to use its good offices to promote an agreement between Iran and Turkey over the Aras River, but no such agreement was reached. The issues concerning the Aras River were quite complex. One complicating factor was the fact that the USSR was also a riparian, indeed a major riparian, of the Aras River. The USSR was not a member of the Bank, and the Bank had no leverage over, or dealings with the USSR at that time.[149] As such, it was not possible for the Bank to approach and effectively involve the USSR. Accordingly, the Bank's

---

[148] As mentioned earlier, the initial date set for consideration of the project by the Bank Executive Directors was June 14, 1983. The project was finally considered and approved by the Executive Directors on June 5, 1984.

[149] The Russian Federation joined the World Bank in 1992, following the collapse of the Union of Soviet Socialist Republics (USSR) in 1991.

efforts in this aspect did not go far.[150] However, Iran did get Bank management to prepare a memorandum on the implementation of its policy for projects on international waters, and to recommend major revisions to the policies and procedures that governed Bank-financed projects on international waterways.

In response to the Executive Directors' discussion and decision on June 5, 1984, a Memorandum entitled "Riparian Rights: The Bank Experience—Staff Review and Recommendations" was prepared by Bank management.[151] The Memorandum was discussed and approved by the Executive Directors on March 7, 1985. The Memorandum included four parts. Part I dealt with the main characteristics of and statistics on water resources, and the global supply and demand situation. Part II surveyed the basic principles of international water law prevailing at that time. Part III dealt with the Bank policy and practice with regard to projects on international waters. Part IV represented the recommendation section of the Memorandum, and included the detailed proposed changes to the then existing policies and procedures (OMS 2.32 of October 1977). Based on the recommendations approved by the Executive Directors in March 1985, a new and more detailed and comprehensive OMS 2.32 was issued in April 1985.

It is worth noting that the first Operational Memorandum for projects on Inland International Waterways (OM 8 of 1956) was issued as a result of the objection of Turkey to the Ghab Project in Syria. The 1985 OMS was issued as result of the objection of Iran to the Igdir-Aksu Project in Turkey and the specific request of Iran that revised guidelines for projects on international waterways be issued. Moreover, as discussed earlier, the few objections that the Bank received regarding some of its financed projects were largely in the Middle East region.[152] This should not come as a surprise. The Middle East region at that time had, and still has, the lowest per capita net water distribution in the world.[153] Although the

---

[150] It could also be argued that the circumstances surrounding the objection, and later the approval of the project by the Executive Directors, were perhaps additional factors for the Bank's efforts not to go far in this aspect.

[151] Memorandum R85-16, IDA/R85-6, dated January 17, 1985. The Committee that prepared the Memorandum was chaired by Dr. Raj Krishna. For other members of the Committee, *see* Krishna, *supra* n. 105, at 31.

[152] *See* Iraq, Lower Khalis Irrigation Project, *supra* n. 141, and Syria, Balikh Irrigation Project, *supra* n. 142.

[153] *See* The World Bank, *Making the Most of Scarcity, Accountability for Better Water Management in the Middle East and North Africa*, at 4 (The World Bank 2007). The Report states that "Per capita renewable water resources in the region, which in 1950 were 4,000m per year, are currently 1,100m per year. Projections indicate that they will drop by half, reaching 550 m per person in 2050. This compares to a global average of 8,900 m per person per year today, and about 6,000 m per person in 2050 when the world population will reach more than 9 billion." *See id.*, at 5.

region is home to five percent of the world's population, it contains less than one percent of the world's annual renewable freshwater.[154] Thus, it should not come as a surprise that the countries of the region have been over-zealous in protecting their interests and claims over the shared rivers in the Region.

The OMS that was issued in April 1985 drew substantially from the principles of international water law, established or emerging, at that time. Before discussion and analysis of the 1985 OMS is undertaken, an overview of those principles of international water law needs to be presented.

3.2 Evolution of International Water Law 1956–1985

The previous Chapter surveyed the basic principles of international water law that had emerged by the middle of the 1950s. The Chapter reviewed the principles governing the navigational uses that started to emerge in the beginning of the nineteenth century, starting with the Final Act of the Congress of Vienna, 1815, which established the principles of freedom and priority of navigation over other uses. The Chapter discussed also the Berlin Act of 1885, and analyzed the reasons for the decline of the supremacy of navigation. The Chapter then reviewed the principles of international water law dealing with non-navigational uses, including the four principles reflecting state practice, the two Conventions adopted under the auspices of the League of Nations (the Barcelona and the Geneva Conventions), the declarations and rules issued by the IIL and the ILA, as well as the judicial and arbitral decisions in this field.

As indicated earlier, the IIL issued its first resolution, the Madrid Declaration, in 1911. The crux of the Declaration is the right of the state to use the waters of an international watercourse in its territory, but only in such manner as not to cause appreciable harm to other riparians. Thus, the Madrid Declaration laid down the basis for the prohibition of causing appreciable harm to other riparians. However, the Declaration required the consent of the other riparians.

It would take 50 years before the IIL would adopt another resolution on international waters. In 1961, the IIL issued the Salzburg Resolution which was referred to in the previous Chapter.[155] The Resolution applies to the utilization of waters which form part of a watercourse or hydrographic basin that extends over the territory of two or more states.[156] The Resolution states that if the riparian

---

[154] *See* The World Bank, *From Scarcity to Security—Averting a Water Crisis in the Middle East and North Africa*, at 5 (The World Bank 1998).

[155] *See* Salzburg Resolution, *supra* n. 69.

[156] *See* Article 1. The Salzburg Resolution continued the use of the term "watercourses" used by the IIL in the Madrid Declaration in 1911. The term would be used again in 1957

states are in disagreement over the scope of their rights of utilization of the shared watercourse, settlement will take place on the basis of equity, taking particular account of their respective needs, as well as of other pertinent circumstances. It addresses notification when it confirms the right of the riparian state to undertake works or utilization of the waters of a watercourse or hydrographic basin that may seriously affect the possibility of utilization of the same waters by other states. Such works can only be undertaken after previous notice is provided to the interested states.[157] In case an objection is made, the Resolution requires the riparian states to enter into negotiations with a view to reaching an agreement within a reasonable period of time. During the negotiations, every state must, in conformity with the principle of good faith, refrain from undertaking the works or utilization that are the object of the dispute, or from taking any other measures that might aggravate the dispute or render agreement more difficult. If the states fail to reach agreement within a reasonable time, the Resolution recommends that they submit the question to judicial settlement or arbitration.[158]

Another noteworthy Resolution of the IIL is on the pollution of rivers and lakes which was adopted in Athens in 1979.[159] The Athens Resolution confirms the sovereign right of the states to exploit their own resources pursuant to their own environmental policies. However, the Resolution subjects the states' right to exploit their own resources to the duty to ensure that their activities will cause no pollution in the waters of international rivers and lakes beyond their boundaries.[160] Any breach of this obligation would result in liability under international law. The Resolution prescribes two means for compliance with this obligation: (i) enactment of necessary laws and regulations, and adoption of efficient and adequate administrative measures and judicial procedures for the enforcement of such laws and regulations; and (ii) cooperation in good faith with the other states concerned.[161]

---

by the Lake Lanoux Tribunal, as will be discussed later. It is worth adding that the term "watercourse" is now widely used in lieu of rivers, drainage basins, or waterways, as will also be discussed later.

[157] *See* Article 5. It is noteworthy that the Article, like the rest of the Resolution, does not deal with upstream and downstream states, but uses the term "interested states."

[158] *See* Articles 5 and 6 of the Salzburg Resolution.

[159] Resolution on "The Pollution of Rivers and Lakes and International Law" adopted by the Institute of International Law, September 12, 1979, 58 Annuaire de l'Institut de Droit International 196 (1979), also known as the "Athens Resolution"; *see* also Harald Hohmann, *Basic Documents of International Environmental Law*, vol. 1, at 256 (1992).

[160] *See id.*, Article II. Article 1 of the Resolution defines pollution as any physical, chemical, or biological alteration in the composition or quality of waters which results directly or indirectly from human action and affects the legitimate uses of such waters, thereby causing injury.

[161] *See id.*, Article IV.

One common feature of the three resolutions is the concern about the effects of the activities undertaken by one riparian state on other riparian states. Unlike the Madrid Declaration, the Salzburg Resolution does not include detailed provisions dealing with the prohibition against causing appreciable harm to other riparians. This issue is addressed by the Salzburg Resolution in connection with works to be undertaken by one state in the shared watercourse which would seriously affect other watercourse states. Here the Resolution, unlike the Madrid Declaration, does not require the consent of other watercourse states for such works. Rather it subjects the matter to notification, negotiations in case of objection, and judicial settlement if the negotiations fail to resolve the issue. Along those lines, the Athens Resolution confirms the right of the states to exploit their own resources, but subjects this right to the duty to ensure that their activities will not cause pollution in the waters of international rivers. Accordingly, it can be concluded that the main feature of the IIL rules and resolution is the obligation not to cause harm to the other riparians.

On the other hand, the work of the ILA on international waters took a different route from that of the IIL. Whereas the IIL addressed international watercourses through the obligation not to cause harm or serious adverse effects, the ILA emphasized the concept of equitable and reasonable utilization. The IIL issued its first declaration in 1911, while the ILA started working on international waters only in 1954. As indicated earlier, the ILA established in that year "the Committee on the Uses of International Rivers." The first resolution to be issued by that Committee was the Dubrovnik Statement, which confirms the rights of every riparian state on the international river but requires these states to exercise those rights with due consideration for the effects on the other riparian states.[162]

The Dubrovnik Statement was followed in 1958 by the New York Resolution.[163] Article 2 of the Resolution confirms the right of each co-riparian state to a reasonable and equitable share in the beneficial uses of the waters of the drainage basin.[164] The principle of reasonable and equitable utilization was the essence of discussions at the Tokyo meeting of the ILA in 1964.[165]

---

[162] *See supra* n. 72.

[163] ILA, Report of the Forty-Eighth Conference 28 (New York 1958) (hereinafter the "New York Resolution").

[164] Article 2 of the New York Resolution stated that a reasonable and equitable share should be determined in light of all the relevant factors of each particular case. However, the Resolution did not include any such factors.

[165] ILA, Report of the Fifty-First Conference 167 (Tokyo 1964). In elaborating on the concept of equitable utilization, the ILA observed: "Any use of water by a riparian State, whether upper or lower, that denies an equitable sharing of uses by a co-riparian State conflicts with the community of interests of all riparian States in obtaining maximum

In 1966, the ILA adopted the Helsinki Rules on the Uses of the Waters of International Rivers.[166] The Rules incorporated and elaborated the previous principles and resolutions adopted successively at Dubrovnik, New York, and Tokyo. Although the title of the Rules refers to international rivers only, Article 1 states that the Rules are applicable to the use of the waters of an international drainage basin. Such a drainage basin is defined as "a geographical area extending over two or more States determined by the watershed limits of the system of waters, including surface and underground waters, flowing into a common terminus."[167] The Helsinki Rules deal extensively with the principle of reasonable and equitable utilization, and have established the principle as the guiding rule of the work of the ILA, as well as of international water law. For that purpose, the Helsinki Rules have specified a number of factors for determining the reasonable and equitable share for each basin state.[168]

---

benefit from the common source. Thus, uses of waters by a riparian State that cause pollution resulting in injury in a co-riparian State must be considered from the overall perspective of what constitutes an equitable utilization." As such, the obligation not to cause significant harm was brought within the realm of the concept of equitable utilization.

[166] ILA, Report of the Fifty-Second Conference 486 (Helsinki 1966) (hereinafter the "Helsinki Rules").

[167] *See* the Helsinki Rules, Article II. It is worth noting that the definition of "drainage basin" includes both "surface and underground waters, flowing into a common terminus." This is the first time that groundwater was included in any resolution or rules of the ILA or IIL. As will be discussed later, this definition includes only groundwater connected to surface water, and does not extend to aquifers that do not contribute water to, or receive water from shared surface waters. For a detailed discussion of transboundary groundwater, *see infra*, Chapter 8 of this Book.

[168] Article V of the Helsinki Rules states that the relevant factors to be considered include, but are not limited to:

(a) the geography of the basin, including in particular the extent of the drainage area in the territory of each basin state;
(b) the hydrology of the basin, including in particular the contribution of water by each basin state;
(c) the climate affecting the basin;
(d) the past utilization of the waters of the basin, including in particular existing utilization;
(e) the economic and social needs of each basin state;
(f) the population dependent on the waters of the basin in each basin state;
(g) the comparative costs of alternative means of satisfying the economic and social needs of each basin state;
(h) the availability of other resources;
(i) the avoidance of unnecessary waste in the utilization of waters of the basin;
(j) the practicability of compensation to one or more of the co-basin states as a means of adjusting conflicts among uses; and
(k) the degree to which the needs of a basin state may be satisfied, without causing substantial injury to a co-basin state.

The Helsinki Rules devote a separate chapter to procedures, not only for settlement but also for the prevention of disputes. These procedures include both the exchange of data and information on the drainage basin between the riparian states, as well as notification. With regard to notification, Article XXIX requires a state, regardless of its location in a drainage basin, to furnish to any other basin state, the interests of which may be substantially affected, notice of any proposed construction or installation which would alter the regime of the basin in a way that might give rise to a dispute. The notice should include such essential facts as would permit the recipient to make an assessment of the probable effect of the proposed alteration. The notifying state should give the recipient a reasonable period of time to make an assessment of the probable effect of the proposed construction or installation and to submit its views thereon to the notifying state. In case of a dispute between states as to their legal rights or other interests, they should seek a solution by negotiation.

The ILA's work on international water law did not taper off after the issuance of the Helsinki Rules. In 1972 the ILA issued its Articles on Flood Control,[169] and in 1976, the Rules on Administration of International Watercourses were adopted.[170] In 1980, the Belgrade Conference of the ILA adopted two sets of rules, known as the "Belgrade Rules." The first set dealt with the regulation of the flow of the water of international watercourses, and the second dealt with the relationship of international water resources to other natural resources' environmental elements.[171] Separate articles on water pollution of an international drainage basin were adopted at the Montreal Conference in 1982.[172] In all of these rules and resolutions of the ILA, the supremacy of the Helsinki Rules has been underscored. Each was subjected to the requirement of being consistent

---

Paragraph 3 of Article V states that the weight to be given to each factor is to be determined by its importance in comparison with that of other relevant factors. *See* Helsinki Rules, *supra* n. 166.

[169] ILA, Report of the Fifty-Fifth Conference 46 (New York 1972).

[170] ILA, Report of the Fifty-Seventh Conference 213–266 (Madrid 1976). The use by the ILA of the term "watercourses" is noteworthy. ILA started using this term synonymously with the term "drainage basin."

[171] ILA, Report of the Fifty-Ninth Conference, 362–373 and 273–375 (Belgrade 1980).

[172] ILA, Report of the Sixtieth Conference (Montreal 1982). The ILA issued a number of other rules in the 1980s. By the late 1980s and early 1990s, it became clear to the ILA that the rules it had adopted were expanding, and provisions governing the same issue may be scattered in more than one instrument. Accordingly, the ILA decided to consolidate those rules in one instrument. The draft of the consolidated rules, known as "The Campione Consolidation of the ILA Rules on International Water Resources, 1966–1999," was completed during the Water Resources Committee meeting in Campione, Italy, in June 1999. Following the issuance of the Campione Consolidated Rules, the ILA decided in

either with the Helsinki Rules or with the principle of equitable utilization, which is itself the foundation of the Rules. Indeed, The Helsinki Rules were regarded soon after their issuance as the most authoritative set of rules for regulating the use and protection of international watercourses. As noted by Charles Bourne, the Helsinki Rules were soon accepted by the international community as reflecting customary international law.[173]

Thus, while the IIL emphasized the obligation not to cause harm, the ILA work centered on the concept of reasonable and equitable utilization. However, it should be emphasized that the IIL and ILA declarations, resolutions and rules have no formal standing or legally binding effects *per se*. Their authoritative effects stem from the expertise and scholarship of those two institutions and their members, and the fact that a number of the provisions of those rules reflect, by and large, customary international water law.

In parallel with the work of the IIL and the ILA, the United Nations General Assembly asked the International Law Commission (ILC)[174] on December 8, 1970, to study the topic of international watercourses with the view of its progressive development and codification.[175] The ILC started working on the draft Convention at its twenty-third session in 1971, and completed its work in 1994. The task was clearly a complex one. It took 23 years, five rapporteurs and 15 reports before the final draft articles of the United Nations Convention on the Law of the Non-Navigational Uses of International Watercourses (Watercourses Convention) were agreed upon by the ILC.[176] The draft Convention was then deliberated by the Sixth Committee of the UN (the Legal Committee), convened

---

2000 to proceed further with the revision and updating of the Helsinki Rules to correspond to the present state of customary international water law. The Committee on Water Resources held a series of meetings and worked on a number of reports on the new rules. The final report which included "the Berlin Rules on Water Resources" was presented and approved by the ILA in Berlin in 2004. *See* ILA, Report of the Seventy-First Conference, 334 (Berlin 2004). One major criticism of the Berlin Rules is their application of international principles to national waters. See the Dissenting Opinion to the Berlin Rules available at: http://www.waterlaw.org/intldocs/ila_berlin_rules_dissent.html. *See* also Salman M. A. Salman, *The Helsinki Rules, the United Nations Watercourses Convention and the Berlin Rules*, 23 International Journal of Water Resources Development 525 (2007).

[173] *See* Charles Bourne, *The International Law Association's Contribution to International Water Resources Law*, *supra* n. 71.

[174] The ILC was established pursuant to the United Nations General Assembly Resolution 174(11) of November 21, 1947, as a subsidiary organ of the General Assembly for the progressive development and codification of international law. The ILC is composed of legal experts nominated by states and elected by the General Assembly.

[175] *See* United Nations General Assembly (UNGA) Resolution 2669 (XXV).

[176] *See* 1994 Yearbook of the International Law Commission, Volume II, Part Two, 88 (1997).

as the Working Group of the Whole.[177] The Working Group, after two lengthy and contentious sessions in 1996 and 1997, adopted the draft Convention with some changes, and recommended it the UNGA on April 4, 1997.[178] It also introduced a number of statements of understanding pertaining to some of the articles of the draft Convention.[179]

Thereafter, on May 21, 1997, following lengthy discussion of the draft articles, as amended by the Working Group, the General Assembly of the United Nations adopted the Convention.[180] Although the Convention has not yet entered into force,[181] there is general agreement among experts in this field that the provisions of the Convention reflect the basic principles of customary international water law.[182] The Convention has also been endorsed by a number of global entities.[183]

---

[177] This was done pursuant to the UNGA Resolution 49/52 of December 9, 1994. The Resolution invited the states to submit written comments and observations on the draft articles by July 1, 1996, and indicated that the Sixth Committee would convene as a Working Group of the Whole, open to States Members of the United Nations, or members of specialized agencies to elaborate a framework convention on the basis of the draft articles adopted by the ILC, and in light of the written comments and observations of states and the debate of the working group.

[178] For some of the contentious issues deliberated by the Working Group, and the agreements reached thereon, *see* Caflisch, *supra* n. 56 at 3.

[179] These statements of understanding covered varied issues, including clarifications to some of the terms used in the draft Convention, as will be discussed later. For the Report of the Six Committee Convening as the Working Group of the Whole, *see*: http://www.un.org/law/cod/watere.htm.

[180] *See* UNGA Resolution 51/229. The Convention was adopted by a vote of 103 for, and 3 against (Burundi, China and Turkey) with 27 abstentions while 52 countries did not participate in the voting. For the full text of the Convention, *see* 36 I.L.M. 700 (1997). *See* also Salman & Boisson de Chazournes (eds.), *surpa* n. 56, Annex 1, at 173.

[181] The Convention needs 35 instruments of ratification or accession to enter into force and effect. As of January 2009, only 16 countries have ratified or acceded to the Convention. For discussion of the status of the Convention, and the reasons of the reluctance of states to become parties to it, *see*, Salman M. A. Salman, *The United Nations Watercourses Convention Ten Years Later—Why Has its Entry into Force Proven Difficult?* 32 Water International 1 (2007).

[182] *See* Stephen McCaffrey, *The UN Convention on the Law of the Non-Navigational Uses of International Watercourses: Prospects and Pitfalls*, in Salman & Boisson de Chazournes (eds.), *supra* n. 56, at 17.

[183] The Convention has been endorsed by a number of entities including the World Water Council (WWC), the World Commission on Water for the 21st Century (WCW), as well as the World Commission on Dams (WCD). *See* Salman M. A. Salman, *Dams, International Rivers, and Riparian States: An Analysis of the Recommendations of the World Commission on Dams*, 16(6) Am. U. Int'l L. Rev. 1477 (2001). *See* also the discussion of the *Gabčíkovo-Nagymaros case* and the endorsement of the International Court of Justice, *infra* n. 326.

The Convention is based largely on the ILA work, particularly the Helsinki Rules which were issued by the ILA in 1966, and to some extent on the work of the IIL. The Convention itself recognizes "the valuable contribution of international organizations, both governmental and non-governmental, to the codification and progressive development of international law in this field."[184] The Convention also recalls the existing bilateral and multilateral agreements regarding the non-navigational uses of international watercourses.

The Convention aims at ensuring the utilization, development, conservation, management, and protection of international watercourses, and promoting optimal and sustainable utilization thereof for present and future generations. The main areas that the Convention addresses include definition of the term "watercourse;" watercourse agreements; equitable and reasonable utilization and the obligation not to cause harm; planned measures; protection, preservation, and management; and dispute settlement. Thus, it is a framework convention that provides basic substantive and procedural principles which subsequent agreements could adopt or adjust.[185]

When the Committee entrusted by Bank management with the task of preparing the Memorandum on Riparian Rights started its work in June 1984, six reports had been issued by the ILC in connection with its work on the draft Watercourses Convention.[186] Although preliminary in nature, those reports, together with the work of the IIL and the ILA, provided major and important guidance for the analysis and recommendations of the Committee to the Bank's Executive Directors regarding the new Bank policy for projects on international waterways, discussed in the next Chapter of this Book.

---

[184] *See* Preamble to the Convention, recitals 9 and 10.

[185] For a detailed analysis of the Convention *see* Attila Tanzi & Maurizio Arcari, *The United Nations Convention on the Law of International Watercourses* (Kluwer Law International 2001). *See* also, Stephen McCaffrey, *The Law of International Watercourses* (2d. ed., Oxford University Press 2007).

[186] The first report was prepared and presented by Mr. Kearney in 1976 (A/CN.4/295); available at: http://untreaty.un.org/ilc/documentation/english/a_cn4_295.pdf. Thereafter, Mr. Schwebel prepared and presented three reports: (i) the first in 1979 (A/CN.41320), *see* Yearbook of International Law Commission, Vol II, Part One (1979), at 143, (ii) the second report in 1980 (A/CN.4/332), *see* Yearbook of International Law Commission, Vol II, Part I (1980), at 159; and (iii) the third report in 1982 (A/CN.4/348) *see* Yearbook of International Law Commission, Vol II, (Part One) (1982) at 65. Mr. Evensen prepared and presented two reports, the first in 1983 (A/CN.4/367), Yearbook of International Law Commission, Vol II (Part One) (1983), at 155 and the second report in 1984 (A/CN. 4/381), Yearbook of International Law Commission, Vol II (Part One) (1984), at 101. For the full list of the reports on the topic prepared by the ILC rapporteurs, *see* Sir Arthur Watts, *The International Law Commission* 1949–98, Volume Two: The Treaties, Part II, 1335 (Oxford University Press 1999).

3.3   The 1985 OMS and the Subsequent Directives

As mentioned in the previous Part of this Chapter, on April 10, 1985, Operational Manual Statement (OMS) 2.32, Projects on International Waterways, was issued.[187] The history of the previous directives, preceding OMS 2.32, is as follows:

(1) OM 8, Projects on International Inland Waterways was issued in 1956. It was replaced by OM 5.05, Projects on International Waters, issued in 1965.
(2) OM 5.05 was replaced in 1971 by OMS 2.22, Projects on International Waters. OMS 2.22 was itself replaced by OMS 2.32, Projects on International Waters, in October 1977.
(3) OMS 2.32, Projects on International Waters of 1977, was replaced in 1985 by OMS 2.32, Projects on International Waterways.

The main elements of the 1985 OMS have continued to govern and regulate the Bank-financed projects on international waterways until now, despite the different titles and structures given to those different policies. Minor changes, largely nomenclature adjustments to reflect organizational changes within the Bank, were introduced, as was done with OM 5.05 of 1965. Those subsequent titles and structures can be listed as follows:

(1) In 1989 OMS 2.32 of 1985 was renamed and reissued as Operational Directive (OD) 7.50, as a result of the Bank gradually changing the names of its operational manual statements into operational directives.
(2) OD 7.50 of 1989 was withdrawn and reissued as OD 7.50 of 1990 after it was decided that the use of the terms "upstream riparians" and "downstream riparians" in the 1989 version was inappropriate.[188] Instead, the 1990 OD used the term "other riparians."[189]
(3) In April 1991 a supplement to OD 7.50 was issued requiring that documentation for any project on international waterways should be accompanied by a map that clearly indicated the international waterway and the location of the project's components.
(4) OD 7.50 was reissued in 1994 in three separate instruments: Operational Policy (OP),[190] Bank Procedures (BP), and Good Practices (GP) 7.50. That

---

[187] A copy of OMS 2.32 (1985) is attached as Appendix 3 to this Book.

[188] *See* Chapter 7, Part 7.2.3 of this Book which discusses insertion of the term "downstream riparian" in the 1994 policy.

[189] A copy of the 1990 OD is attached as Appendix 4 to this Book.

[190] Operational Policies (OPs) are short, focused statements that follow from the Bank's Articles of Agreement, its General Conditions applicable to loans and credits, and from

action reflected the decision to separate the substantive policy elements from both procedural ones and good practices, in all Bank operational directives, and to include each of them in a separate instrument.[191] In addition, the new OP 7.50 included a third exception to the notification requirement dealing with projects in a tributary that originates, and runs exclusively in the lowest downstream riparian. This exception will be discussed in details later.[192]

(5) OP/BP/GP 7.50 of 1994 were reissued in 2001 to reflect organizational changes within the Bank. The 2001 versions of the OP and BP 7.50 are still the current ones.[193]

In 1997, the Bank carved out 10 policies from its vast body of operational policies and placed them under the umbrella term "the safeguard policies." The Bank policy for projects on international waterways is one these 10 safeguard policies.[194] The Bank decided that these policies are critical to ensuring that potentially adverse environmental and social impacts of Bank-financed projects are identified, minimized, and mitigated. Because of those concerns, the safeguard

---

policies specifically approved by the Executive Directors. The General Conditions are incorporated in the Loan or Credit Agreements.

[191] Good Practices (GP) were soon discontinued and a number of them have been replaced with sourcebooks, guidebooks and handbooks. Examples include the Environmental Assessment Sourcebook, Involuntary Resettlement Sourcebook, Pest Management Guidebook, and Forests Sourcebook. For a list of these sourcebooks *see*: http://web.worldbank.org/WBSITE/EXTERNAL/PROJECTS/EXTPOLICIES/EXTSA FEPOL/0,,contentMDK:20583835~menuPK:1430924~pagePK:64168445~piPK:6416 8309~theSitePK:584435,00.html. However, GP 7.50 has not yet been replaced by a sourcebook.

[192] As will be discussed later, the new paragraph of the OP used the term "downstream riparian." As mentioned in this Part of this Book, OD 7.50 of 1989 was reissued in 1990 solely to replace the terms "upstream riparian" and "downstream riparian" with the term "other riparians."

[193] Copies of the 2001 OP and BP 7.50 are attached as Appendix 5A and 5B, respectively, to this Book. Although GP 7.50 is not included in the current Operational Manual of the Bank, a copy of this GP is attached as Appendix 5C to this Book. It should be noted that the responsibility for overseeing the interpretation and application of this Policy, as well as the Policy for Projects in Disputed Areas (*infra* Chapter 9, Part 9.2 of this Book), rests with the Environmental and International Law Unit of the Legal Vice Presidency (LEGEN) of the World Bank. This is because the issues involved in both policies relate to international law.

[194] The 10 safeguard policies are: (i) Environmental Assessment (OP/BP 4.01); (ii) Natural Habitats (OP/BP 4.04); (iii) Forests (OP/BP 4.36); (iv) Pest Management (OP 4.09); (v) Safety of Dams (OP/BP 4.37); (vi) Physical Cultural Resources (OP/BP 4.11); (vii) Involuntary Resettlement (OP/BP 4.12); (viii) Indigenous Peoples (OP/BP 410); (ix) Projects on International Waterways (OP/BP 7.50); and (x) Projects in Disputed Areas (OP/BP 7.60). In addition, the Bank Policy on Disclosure of Information

policies receive particular attention during the project design, preparation, approval, and implementation, and utmost due diligence is exercised to ensure full compliance by the Bank with those policies.[195]

The basic elements of the Bank policy for projects on international waterways, included in OMS 2.32 of 1985, remain largely the same until today, despite the different titles given to each instrument. The current version of the policy, OP/BP/GP 7.50 of 2001 includes, more or less, the same elements of OMS 2.32 of 1985. The only difference is that those elements have been segregated into policy, procedures and good practice and included in three separate instruments. Accordingly, the discussion in the next chapters will deal with the main elements of the Bank policy as reflected in the 2001 OP/BP/GP, and not with each instrument issued after 1985.

3.4 Approaches of other International Financial Institutions

It is worth adding that the current policy for projects on international waterways applies only to the World Bank (IBRD and IDA), and not to the other members of the World Bank Group (IFC, MIGA, and ICSID). Each of the IFC[196] and MIGA[197] has its own set of operational policies and procedures, and the funding of projects is not relevant to ICSID's mandate.[198]

(2002) is usually associated with the safeguard policies because of the basic requirements under most of the safeguard policies for disclosure of certain assessments and plans, and consultation on such plans. For more information on the safeguard policies *see*: http://web.worldbank.org/WBSITE/EXTERNAL/PROJECTS/EXTPOLICIES/EXTSAF EPOL/0,,menuPK:584441~pagePK:64168427~piPK:64168435~theSitePK:584435,00. html.

[195] As will be discussed later, *see infra*, Chapter 9, Part 9.3 of this Book, the safeguard policies constitute the bulk of the requests for inspection of Bank-financed projects submitted to the Inspection Panel.

[196] IFC is the World Bank Group entity with a mandate to invest in private sector projects in developing member countries. It lends directly to, and makes equity investments in, private companies without guarantees from governments and attracts other sources of funds for these projects. IFC was established in 1956, and as of January 2009, it had 181 member countries. For more information on IFC *see*: http://www.ifc.org/.

[197] MIGA is the World Bank Group entity with the mandate of providing political risk insurance for foreign investments in developing member countries. MIGA was established in 1988, and as of January 2009, it had 173 member countries. For more information on MIGA *see*: http://www.miga.org.

[198] ICSID was established under the Convention on the Settlement of Investment Disputes between States and Nationals of Other States. ICSID provides facilities for conciliation

The IFC adopted its current environmental and social policies in 2006,[199] replacing a number of operational polices in that field that were in effect until that time. The new policies, termed "Performance Standards (PS)," address, to a large extent, similar issues as those addressed by the World Bank safeguard policies.[200] The IFC Policy on Social and Environmental Sustainability put the PS into effect, and Guidance Notes (GN), corresponding to each PS, have been issued to provide implementation guidance to IFC clients.[201] IFC "Environment and Social Review Procedure" provides implementation guidance to IFC staff, in a manner similar to the Bank Procedures.

Projects on international waterways were governed until 2006 by IFC Operational Policy 7.50, which was based largely on the Bank policy.[202] IFC Operational Policy 7.50 is no longer in effect, and IFC projects on international waterways are now covered under its Policy on Social and Environmental Sustainability. This Policy states that IFC has a liaison role to provide "formal notification to countries affected by the transboundary effects of proposed project activities, to help those countries determine whether the proposed project has the potential for causing adverse effects through air pollution or deprivation of water from or pollution of international waterways."[203] The Policy in turn refers

---

and arbitration of investment disputes between contracting states and nationals of other contracting states. ICSID's Convention entered into force on October 14, 1966, and as of January 2009, ICSID had 143 members. For the ICSID's Convention, and for more information on ICSID, *see*: http://icsid.worldbank.org/ICSID/ICSID/RulesMain.jsp.

[199] A description of those policies is available at: http://www.ifc.org/ifcext/sustainability. nsf/Content/EnvSocStandards.

[200] IFC has eight Performance Standards (PS): PS1, Social and Environmental Assessment and Management System; PS2, Labor and Working Conditions; PS3, Pollution Prevention and Abatement; PS4, Community Health, Safety and Security; PS5, Land Acquisition and Involuntary Resettlement; PS6, Biodiversity Conservation and Sustainable Natural Resources Management; PS7, Indigenous Peoples; and PS8, Cultural Heritage. Those PS are used by the commercial banks that have adopted what is referred to as the Equator Principles. The Equator Principles are "a financial industry benchmark for determining, assessing and managing social and environmental risks in project financing." For more information on the Equator Principles, *see*: http://www.equator-principles.com/index.shtml.

[201] A copy of the Policy on Social and Environmental Sustainability is available at: http:// www.ifc.org/ifcext/sustainability.nsf/AttachmentsByTitle/pol_SocEnvSustainability2006/ $FILE/SustainabilityPolicy.pdf. For the Guidance Notes, *see supra* n. 199.

[202] That Operational Policy was issued in November 1998. One main difference from the Bank policy was that, for the purposes of notification, IFC may request the assistance of the relevant members of IFC's Board of Directors.

[203] *See* paragraph 40 of the Policy on Social and Environmental Sustainability (April 30, 2006), *supra* n. 201. Moreover, paragraph 6 of the Policy on Environmental Assessment

to the IFC Performance Standards. Guidance Note 1 to PS1 includes the same definition of the term "international waterways" as that set forth in the Bank policy,[204] and restates IFC's willingness to assist the client with the notification process if requested by such a client. Basic steps for notification are included in the IFC "Environment and Social Review Procedure." Such steps include notification through the members of IFC Board of Directors, and identification of the need for the appointment of one or more external experts to review and analyze the transboundary impacts of the proposed project.

In October 2007, MIGA adopted Performance Standards and Guidance Notes similar to those of IFC, including the definitions and procedures set therein.[205]

It is worth noting that none of the major regional financial institutions has a separate policy for projects on international waterways. The European Bank for Reconstruction and Development (EBRD) Environmental and Social Policy adopted in 2008 requires that the appraisal of the environmental and social impacts of the project should ". . . also consider potential transboundary and global issues, such as impacts from effluents and emissions, increased use or contamination of international waterways, greenhouse gas emissions, climate change mitigation and adaptation issues, and impacts on endangered species and habitats."[206] The EBRD's Policy does not include any other requirements for dealing with projects on international waterways.

The Inter-American Development Bank (IDB) goes a step further and requires notification of affected countries, as well as consultation with affected parties.

---

requires that the ". . . The Assessment will also consider potential transboundary effects, such as pollution of air, or use or pollution of international waterways, as well as global impacts, such as the emission of greenhouse gasses . . . ."

[204] *See* paragraph G21, n. 3, of Guidance Note 1, Social and Environmental Assessment and Management Systems, *supra* n. 199.

[205] For MIGA's Performance Standards, *see*: http://www.miga.org/policies/index_sv.cfm?stid=1652.

[206] *See* paragraph 7 of EBRD Performance Requirement 1, Environmental and Social Appraisal and Management, available at: http://www.ebrd.com/about/policies/enviro/policy/review/index.htm. It should be added that most of the members of the EBRD are parties to a number of conventions that deal with transboundary impacts such as the Aarhus Convention (Convention on Access to Information, Public Participation in Decision-making, and Access to Justice in Environmental Matters), as well as the Espoo and Helsinki Conventions, and thus they would be bound by the requirements for notification and exchange of data and information under those Conventions. The activities listed in Appendix 1 to the Espoo Convention are classified under EBRD's Policy as Category A projects, requiring "formalized and participatory assessment process carried out by third party specialist." However, no requirement of notification of other riparians is specified under the EBRD Policy, perhaps because notification is already required under the Espoo Convention.

The IDB's Environment and Safeguard Compliance Policy requires that "The environmental assessment process for operations with potentially significant transboundary environmental and associated social impacts, such as operations affecting another country's use of waterways, watersheds, coastal marine resources, biological corridors, regional air sheds and aquifers, will address the following issues: (i) notification to the affected country or countries of the critical transboundary impacts; (ii) implementation of an appropriate framework for consultation of affected parties; and (iii) appropriate environmental mitigation and/or monitoring measures, to the Bank's satisfaction."[207] However, no procedures for notification are included in the IDB's Environmental Policy.

Neither the Asian Development Bank (ADB), nor the African Development Bank (AfDB) has a specific operational policy for projects on international waterways. The ADB Water Policy,[208] and the AfDB Policy for Integrated Water Resources Management,[209] both address international waters in a general way, and underscore the need for cooperation amongst the riparians, and for the establishment and strengthening of river basin organizations. However, these are policy papers, and not operational policies with strict compliance requirements, particularly that of notification of other riparian states.

Thus, the World Bank (IBRD and IDA) stands out as the only international financial institution with detailed polices and procedures for projects on international waterways. Although the IFC and MIGA now have their own policies, it is likely that the general provisions included therein will be complemented by the detailed provisions as well as the precedents and practice of the World Bank. It should, however, be clarified in this respect that this Book deals only with the policy for projects on international waterways of the World Bank, and does not discuss or analyze those of the IFC or MIGA.

---

[207] *See* paragraph B.8, Transboundary Impacts, IDB Environment and Safeguards Compliance Policy, available at: http://www.iadb.org/sds/ENV/publication/publication_183_3923_e.htm.

[208] *See* Asian Development Bank, *Water for All: The Water Policy of the Asian Development Bank* (ADB 2001), available at: http://www.adb.org/Documents/Policies/Water/default.asp.

[209] *See* African Development Bank & African Development Fund, Policy for Integrated Water Resources Management, April 2000. For more information on water resources management at the AfDB *see*: http://www.afdb.org/portal/page?_pageid=533,8250449&_dad=portal&_schema=PORTAL.

CHAPTER 4

Main Features of the Current Bank Policy

4.1   The Pioneering Nature of the Policy

The issuance of OMS 2.32 in 1985 completed a lengthy and complex process that aimed at finding ways for the Bank to handle projects on international waterways. That process spanned over a period of more than 30 years and passed through three stages. The first one commenced shortly after the Bank began financing projects that involved international rivers. The Bank started learning by experience, and gradually developed some basic approaches. Such approaches included not financing projects on rivers where a dispute existed, or those which would cause appreciable harm to other riparians without their consent. However, those "negative" commandments had to be supplemented by some "positive" rules. Some of the approaches deliberated included regional planning, recognition of established uses, and requiring the consent of all the riparians. However, none of those approaches was found practical. Finally the flexible or *ad hoc* approach was adopted. Under this approach, the circumstances of each project and waterway would determine how the Bank would handle such a project. There might be an agreement in place, or the interests of other riparians might not be affected. Similarly other riparians might have no objection to the project, or the objection could be without merit.

As a result of the deliberations on the Ghab Project in the 1950s, OM 8 was issued in 1956, establishing an early warning system. Nine years later, OM 5.05 codified some of the above practices, and reiterated the early warning system procedures of OM 8.

Gradually, gaps and ambiguities in the 1965 OM started to surface. This situation was compounded by the growth and increasing novelty in Bank-financed projects, particularly after the establishment of IDA in 1960,[210] and the expansion of the membership of both IBRD and IDA.[211] Escalation in the competing demands of the different riparians over their shared waterways, particularly in the

---

[210] As mentioned earlier, IDA is the arm of the World Bank that helps the world's poorest countries through interest-free credits. *See supra* n. 119.

[211] When OMS 2.32 was issued in 1985, IBRD had a total membership of 148 countries, while IDA had 133 member countries; *see The World Bank Annual Report* 1985, at 30. Membership of IBRD grew considerably over the years, from 38 in 1946, to 58 in 1956, to 103 in 1965, and to 148 in 1985; *see supra* n.118.

Middle East, was another factor that prompted a rethinking in the Bank of the modalities for handling projects on international waterways.

The paucity of international law rules in this field was clear. There was no universal treaty in force regulating the uses of international watercourses in 1985, and indeed, there is still no such treaty even today. The ILC work on the draft Watercourses Convention was, in 1985, still a long way from completion. And although the IIL and the ILA had issued some rules, declarations and resolutions in this field, these rules, *per se*, had no legally binding effects. Keeping all these factors in mind, OMS 2.32 of 1985 was without doubt a pioneering instrument.[212] It incorporated and reflected the Bank's extensive experience of more than 35 years of work on projects on international waterways, during which the previous OMs had been tested. It was the first such instrument in the field of international waters for which there was a requirement for compliance by the Bank and its borrowers at an almost global level. No other similar instrument existed at that time, nor does any one exist today. The provisions of the policy either withstood the test of global application and time, or were adapted to fit the new circumstances and challenges. The fact that the policy parameters of the Bank are still largely those of 1985 testifies to this fact.

Moreover, OMS 2.32 of 1985, as well as each of its subsequent reiterations, (hereinafter "the policy"), is a fairly comprehensive instrument. The policy sets out the objectives that it aims to achieve, centering mainly on riparian cooperation and the need to notify the other riparians of Bank-financed projects. It deals with the definition of international waterways as well as the projects covered thereunder. The policy then describes the internal process for addressing potential international water rights issues, and thereafter sets forth in great details the process for notification of other riparians (by the borrower or the Bank), and how the Bank would handle the responses to such notification. To ensure the transparency of the process, the policy specifies the information regarding the international waterways aspects of the project that needs to be included in the project documents (initially called the Staff Appraisal Report (SAR), and later changed to the Project Appraisal Document (PAD)). The policy includes detailed procedures for obtaining an additional opinion of independent experts on objections received from one or more of the riparians affected by the project.

Another remarkable feature of the policy is its reach beyond projects and notification. It states that the Bank attaches utmost importance to the riparians entering into collaborative arrangements or agreements for the entire waterway or any

---

[212] As discussed earlier, OMS 2.32 of 1985 was preceded by OM 8 of 1956 and OM of 1965, but those two memoranda, as we have seen, were limited in scope, and were not widely known outside the Bank, perhaps because they did not pronounce any substantive rules.

part thereof, and announces the Bank's readiness to assist the riparians to this end. Each of these areas is discussed and analyzed in more detail below.

4.2 Objectives of the Policy

The objective set forth under each of the 1956 and 1965 OM was quite limited. Each instrument stated that projects on international waterways could affect relations not only between the Bank and the borrower, but also between governments, and because of that, such projects need special handling.[213] It is basically a dispute avoidance approach. That objective has been refined and considerably expanded under the policy in a number of ways.

The policy emphasizes the fact that the cooperation and goodwill of the riparians is essential for the efficient utilization and exploitation of international waterways for development purposes.[214] It may be recalled that one of the early approaches to shared rivers deliberated by the Bank was regional planning. Such regional planning would be carried out through the establishment of a river basin management authority to study and determine the most effective ways for the utilization of the shared river. Rather than mandating it as an approach for dealing with international waterways, as was deliberated earlier, the Bank policy sets this as an objective which the Bank will assist the riparians in attaining.

The policy goes on to assert that the Bank attaches the utmost importance to riparians' entering into agreements or arrangements for such utilization for the entire waterway or any part thereof,[215] and to declare the readiness of the Bank to assist the riparians to this end. This is a clear pronouncement by the Bank of its willingness to engage in a facilitative role in the intricate and controversial arena of international waters, and to assist in dispute prevention as well as resolution.

One basin with which the Bank's name has been closely associated is the Indus River System, where the Bank played an extraordinary role in the resolution of

---

[213] *See* paragraph 1 of OM 8, and paragraph 1 of OM 5.05.

[214] *See* paragraph 3 of OP 7.50.

[215] The phrase "or any part thereof" should not be understood as an encouragement by the Bank policy for the parceling of the basin into parts. There are cases where parts of a large basin can be treated and dealt with separately, albeit still within the overall basin. The Sava Basin is one such example. The Sava River is tributary of the Danube. Nonetheless, the Framework Agreement on the Sava River Basin was concluded by the four riparian countries (Slovenia, Croatia, Bosnia and Herzegovina, and Serbia) on December 3, 2002, and entered into force on December 29, 2004. The Agreement establishes the International Sava River Basin Commission (Sava Commission). However, the Agreement reiterates the fact that the Sava River Basin is part of the Danube Basin, and calls for coordination of the activities of the Sava Commission with the International Commission for Protection of Danube River (ICPDR); *see*: http://www.savacommission.org/history. For more details on the Danube River, the International Commission for the Protection of the Danube River (ICPDR), and the Danube Convention *see supra* n. 413.

the Indus dispute between India and Pakistan.[216] Indeed, the Bank's role there started even before the issuance of its first policy on international waters in 1956. The dispute erupted as a result of the partitioning of the Sub-continent into India and Pakistan. The borders between the two countries were drawn across the Indus River, cutting the Indus irrigation system into two. Most of the water-rich headwaters went to India, and Pakistan was left as the water-short lower riparian. The dispute escalated and threatened peace and stability in the region. As mentioned earlier, because of the dispute, the Bank was not able to finance the Bhakra-Nagal Project in India, or the Lower Sind Barrage in Pakistan.[217] The then President of the Bank, Eugene Black, wrote to the two Prime Ministers of India and Pakistan on September 6, 1951, offering the Bank's "good offices," an offer which both countries readily accepted.[218] It took nine years of intensive and complex negotiations, with the active mediation of the Bank, before a treaty was concluded and signed by the two parties and the World Bank.[219]

The Treaty is a comprehensive, elaborate, and complex instrument spanning more than 150 pages, and divided into the main part and eight annexures. Some of the annexures include appendices attached thereto. The Treaty basically divided the six rivers comprising the Indus System into two groups, the Eastern River and the Western River, and allocated the former to India, and the latter to Pakistan.[220] Notwithstanding this division, each of the two parties was allowed certain uses of the rivers allocated to the other.[221] The Treaty established the Permanent Indus Commission for overseeing implementation of the Treaty, and for

---

[216] For a detailed analysis of the Indus dispute, *see* Niranjan D. Gulhati, *Indus Waters Treaty—An Exercise in International Mediation* (Allied Publishers 1973). *See* also Aloys Arthur Michel, *The Indus River—A Study of the Effects of Partition* (Yale University Press 1967). *See* also Salman & Uprety, *supra* n. 61.

[217] *See supra*, Chapter 2, Part 2.1 of the Book.

[218] Intervention of the World Bank came as a result of an article by Mr. David Lilienthal, former Chairman of the Tennessee Valley Authority (TVA), published in August 1951. The article, entitled "Another Korea in the Making?" (5 Collier's, July–Sept 1951, at 22), suggested that the World Bank was the best suited institution to play a role in resolving the Indus River System dispute. The Bank took up that suggestion.

[219] *See* The Indus Waters Treaty 1960 Between the Government of India, the Government of Pakistan and the International Bank for Reconstruction and Development, signed at Karachi, Pakistan, on September 19, 1960; 419 U.N.T.S. 126. Prime Minister Jawaharlal Nehru signed on behalf of India, President Mohammad Ayub Khan on behalf of Pakistan, and Mr. William A. Iliff on behalf of the Bank.

[220] India was allocated the Eastern Rivers (the Sutlej, the Beas and the Ravi), and Pakistan the Western Rivers (the Indus, the Jhelum and the Chenab).

[221] Annexure B of the Treaty deals with agricultural use by Pakistan of certain tributaries of the Ravi River, which has been allocated to India. On the other hand, Annexure C of the Treaty deals with agricultural use by India of the Western Rivers allocated to Pakistan.

resolving questions arising therefrom.[222] The Treaty includes detailed provisions for settling issues that may arise in the implementation of the Treaty, which cannot be resolved by the Commission, or which are outside the jurisdiction of the Commission, as will be discussed below.

A number of reasons contributed to the success of the mediation efforts of the Bank.[223] One main reason was the continued and active involvement of the Bank at the highest level. President Black corresponded regularly with the two Prime Ministers and visited both countries three times during the negotiation process. He appointed his assistant, Mr. William Iliff, as the chief mediator, and to ensure continuity, he kept Mr. Iliff in that position even after he was appointed vice president. As a result of that continued and high level involvement, the Bank was able to obtain concessions from the two parties, and to apply pressure as negotiations got difficult and complex. The ability of the Bank to use its leverage and convening power was a major factor in the successful raising of the funds needed for carrying out the works required for implementation of the Treaty.[224] Such funds, which were largely grants, totaled about US$800 million, and were administered by the Bank through the Indus Basin Development Fund.[225] More importantly,

---

Moreover, Annexure D addresses generation of hydroelectric power by India on the Western Rivers, and Annexure E deals with storage of waters by India on the Western Rivers. It is noteworthy that India has been allowed more uses of the Western Rivers than Pakistan's of the Eastern Rivers, because it is the upper riparian, and those rivers flow for long distances in India before entering Pakistan. However, the provisions of Annexures C, D, and E lay down detailed conditions for the uses of the Western Rivers by India.

[222] *See* Article VIII of the Treaty. The Commission consists of one Commissioner representing each party. The Commission is entrusted, *inter alia*, with undertaking every five years a general tour of inspection of all the rivers for ascertaining the facts connected with various developments and works on those rivers. Moreover, the Commission acts as a conduit for exchange of data and information and for undertaking notification for the purposes specified under the Treaty. However, the Commission has no legal status, and does not have a headquarters or secretariat, as do most other river basin commissions.

[223] For discussion of the reasons behind the Bank's success, *see* Salman M. A. Salman, *Good Offices, Mediation and International Water Disputes*, in *Resolution of International Water Disputes* 155 (The International Bureau of the Permanent Court of Arbitration (ed.), Kluwer Law International 2003).

[224] Works were needed to augment the flow of the Western Rivers so as to make the water available to Pakistan close to the pre-Treaty levels, and to end Pakistan's reliance on the Eastern Rivers. Such works included, *inter alia*, two storage dams, eight link canals nearly 400 miles long, and 2,500 tubewells. Those works were called "replacement works" by the Treaty, to distinguish them from other development works that are incremental in nature.

[225] The countries that provided grant funding to assist in the financing of the replacement works were Australia (AUD6,965,000); Canada (CAD22,100,000); Germany (DM126 million); United Kingdom (GBP20,860,000); and the United States (US$177 million). India contributed US$174 million. The Bank provided a loan of US$90 million. Those funds were managed by the Bank as the "Indus Basin Development Fund."

the Bank is a signatory to the Treaty for certain specified purposes. Indeed, this is the only international waters treaty that is signed by a third party.

The purposes for which the Bank signed the Treaty are specified in Articles V and X, and Annexures F, G and H of the Treaty. Article V deals with the financial provisions, including the establishment and administration of the Indus Basin Development Fund which consisted of the funds referred to above.[226] Article X deals with emergency situations that might have interfered with completion of the works funded under the Indus Basin Development Fund. Annexure F deals with the Neutral Expert to be appointed to resolve differences between the two parties. Annexure G deals with the Court of Arbitration to be established to resolve disputes between them. Annexure H deals with the arrangements during the transitional period when Pakistan was discontinuing its reliance on the Eastern Rivers. The responsibilities of the Bank under Articles V and X, and Annexure H were completed in the 1970s. The only remaining responsibilities of the Bank relate to settlement of differences and disputes under Annexures F and G.[227]

The Treaty has established a very unique system for resolving issues that may arise between the two parties. Questions regarding interpretation and application of the Treaty are referred to and decided by the Permanent Indus Commission. If the Commission fails to resolve any such question, then it becomes a difference. Differences are dealt with by the Neutral Expert, and disputes by the Court of Arbitration. The Treaty specifies in Annexure F the questions to be referred to the Neutral Expert (which would then automatically become "differences"). Questions not specified in Annexure F constitute disputes that can only be referred to, and resolved by the Court of Arbitration.

If the parties fail to agree on a Neutral Expert, and also fail to agree on a third party to appoint such a Neutral Expert, then the Bank would have to do so after consultation with the two parties. The decision of the Neutral Expert is final and binding and not appealable to the Court of Arbitration. The Bank manages a trust fund established under the Treaty (Indus Basin Trust Fund) to defray the cost of the Neutral Expert.[228] The Bank also has a role in the selection of the members of

---

[226] In 1962 additional funds were added by the participating governments to the Indus Basin Development Fund when it was reported that the original funds were not sufficient to complete the agreed upon replacement works. In May 1968 the Bank set up the Tarbela Development Fund to supplement the funds allocated for the Tarbela Dam under the Indus Basin Development Fund. *See* Mason & Asher, *supra* n. 2, at 626–27.

[227] Article IX of the Treaty sets forth the main procedures for settlement of differences and disputes. The details of those procedures are laid down in Annexures F and G.

[228] Each party was required by the Treaty to pay $5,000 for the establishment of this trust fund, which the two parties did in 1960. By the year 2005, the amounts in the trust fund exceeded $100,000 due to the investment of the original $10,000 paid by the two parties in 1960. The Indus Basin Trust Fund should be distinguished from the Indus Basin Development Fund established to finance the replacement works in Pakistan, *see supra* n.225.

the Court of Arbitration to be established under Annexure G of the Treaty. Thus, the structure of the process for the settlement of issues by the three bodies (Commission, Neutral Expert, and Court of Arbitration) is not hierarchical, but rather relates to the jurisdiction of each of them.

Clearly, the Bank played a major and constructive role in the conclusion and implementation of the Treaty. This role has been described as "proactive, neutral, pragmatic and fair."[229]

The Bank continues to be involved with the Indus Basin, as per the provisions of the Indus Waters Treaty to which the Bank is a party. The Bank was called upon in 2005 to exercise its responsibilities under Annexure F of the Treaty in connection with the Baglihar difference. This was the first time since the Treaty was concluded in 1960 that the Bank became directly involved in a difference or dispute between the two parties over the Indus Basin. On January 15, 2005, Pakistan approached the World Bank, asking the Bank to appoint a Neutral Expert to address a "difference" that had arisen with India under the Indus Waters Treaty. The difference related to the Baglihar hydropower plant which was, at that time, under construction by India on the Chenab River allocated to Pakistan. The Bank appointed a Neutral Expert four months later, after lengthy consultations with the two parties.[230] It then oversaw the process of the resolution of the difference by the Neutral Expert, and used the Indus Basin Trust Fund to cover its cost.[231] On February 12, 2007, about 20 months after his appointment, the Neutral Expert issued his decision on the difference. The outcome was accepted by the two parties. The process and outcome attested to the strength and credibility of the Treaty, and underscored the importance of the role played by the Bank.[232]

This type of proactive, facilitative role of the Bank is referred to in other Bank instruments as well. The World Bank Water Resources Management Policy Paper

---

[229] *See* Syed Kirmani & Guy Le Moigne, *Fostering Riparian Cooperation in International River Basins: The World Bank at its Best in Development Diplomacy*, World Bank Technical Paper No. 335, at 3 (The World Bank 1997).

[230] For a detailed description and analysis of the Baglihar process *see* Salman M. A. Salman, *The Baglihar Difference and its Resolution Process—A Triumph for the Indus Waters Treaty?* 10 Water Policy, 105 (2008).

[231] At the request of the Neutral Expert, the Bank designated ICSID to coordinate the process, including communication with the two parties. Both parties endorsed this designation. ICSID played a significant coordinating and administrative role, drawing on its rich experience in the settlement of investment disputes, while the Bank continued to oversee the whole process. It is worth noting that only Pakistan is a party to the ICSID Convention. India has neither signed, nor acceded to the Convention. For the role of ICSID in the process, *see* Salman, *id.,* at 111.

[232] The Neutral Expert is mandated under the Treaty to decide which party should bear the cost of the process. In this case he decided that the cost should be borne equally by the two parties. This was done, and the Indus Basin Trust Fund reverted at the end of the process to its 2005 level.

elaborates the areas where the Bank assistance to riparian countries may be needed and is appropriate. After referring to the policy for projects on international waterways, the Policy Paper states that:

> The Bank will help countries improve their management of shared international water resources, for example, by supporting the analysis of development opportunities forgone because of international water disputes. Through technical, financial, and legal assistance, the Bank, if requested, will help governments establish or strengthen institutions, such as river basin organizations, to address transnational water management activities. Furthermore, the Bank will support studies and consultations to review available organizational arrangements and help develop alternative solutions. A flexible approach will be adopted in any initial contact with riparians, avoiding preconditions to the extent possible, in order to explore the most appropriate form of assistance that the Bank may offer. The Bank will be sensitive at all times to the interests of all riparian parties, as all parties must be treated even-handedly. The focus will be on international watercourses in which the Bank's assistance is likely to have a substantial impact.[233]

In connection with the aforementioned objectives and strategy, mention should also be made of the Nile Basin Initiative (NBI).[234] The NBI was established by the Nile Council of Ministers of Water Affairs in 1999, bringing together for the first time the 10 riparian states.[235] It is guided by a shared vision

---

[233] *See* The World Bank, *Water Resources Management—A World Bank Policy Paper*, at 75 (The World Bank 1993), available at: http://web.worldbank.org/WBSITE/EXTERNAL/ TOPICS/EXTWAT/0,,contentMDK:21641718~menuPK:4602418~pagePK:210058~piP K:210062~theSitePK:4602123,00.html. Although issued in 1993, the Policy Paper still provides overall guidance for the engagement of the World Bank on water resources management. It has been complemented in 2003 by the *Water Resources Sector Strategy— Strategic Directions for World Bank Engagement.* The Strategy emphasizes, *inter alia*, that "cooperation on international waters can provide a vital component for broad-based economic development and regional security. A number of the largest water management interventions by the World Bank, dating back to the Indus Waters Treaty of 1960 and extending forward to current projects (including the Lesotho Highlands Water Project and regional initiatives for the Mekong and Nile) fall into this category." *See id.*, at 8. The Strategy is available at: http://web.worldbank.org/WBSITE/EXTERNAL/COUNTRIES/SOUTHASIA EXT/EXTSAREGTOPWATRES/0,,contentMDK:20729817~isCURL:Y~menuPK:494 266~pagePK:34004173~piPK:34003707~theSitePK:494236,00.html.

[234] The Nile 10 riparian states are Burundi, the Democratic Republic of Congo (DRC), Egypt, Eritrea, Ethiopia, Kenya, Rwanda, Sudan, Tanzania and Uganda. For a detailed analysis of the political geography of the Nile *see* Robert O. Collins, *The Nile* (Yale University Press 2002). *See also* John Waterbury, *The Nile Basin—National Determinants and Collective Action* (Yale University Press 2000).

[235] The NBI was officially established by the Nile Basin States at the meeting of their Ministers of Water Resources (or Water Affairs), held in Dar-es-Salaam, Tanzania, on

"to achieve sustainable socio-economic development through equitable utilization of, and benefit from, the common Nile Basin water resources."[236] The NBI is managed by a transitional institutional structure, including the NBI Secretariat located in Entebbe,[237] and project offices in Addis Ababa, and Kigali.[238] The Nile Trust Fund has been established with funding from a number of multilateral and bilateral donors, and some projects are already being funded therefrom. It should be noted, however, that the Nile riparian states have not yet succeeded in concluding a treaty encompassing all the riparian states for the equitable utilization and management of the Nile Basin.

The interventions of the Bank in these two basins, the Indus and the Nile, are the most widely known of the Bank's involvements in international watercourses. In the case of the Indus Basin, a complex and difficult dispute existed, and the Bank played a vital role in its resolution and the conclusion of the Indus Waters Treaty. The Treaty includes provisions for addressing future differences and disputes, and those provisions have proven their vitality in the resolution of the Baglihar difference. On the other hand the role of the Bank in the Nile Basin has been a facilitative one. The NBI aims at bringing the 10 Nile riparians together to share equitably the benefits of the Basin. In other words, there is no specific dispute to be resolved, as was the case with the Indus. The larger aim of the NBI

February 22, 1999, as a transitional arrangement to foster cooperation and sustainable development of the Nile River for the benefit of the inhabitants of those countries. The Nile Basin Initiative Act was enacted by the Republic of Uganda in October 2002 to give the force of law in Uganda to the signed Agreed Minute No. 7 of the 9th Annual Meeting of the Nile Basin States' Ministers of Water Affairs, held in Cairo, Egypt, on February 14, 2002. According to the Agreed Minute, the Ministers decided to "invest the NBI, on a transitional basis, with legal personality to perform all of the functions entrusted to it, including the power to sue and be sued, and to acquire or dispose of moveable and immoveable property." For more details on the NBI, including those aspects, *see*: http://www.nilebasin.org/index.php?option=com_content&task=view&id=13. It should be clarified that Eritrea is not yet a full-fledged member of the NBI, and only holds observer status.

[236] *See id. See* also, International Consortium for Cooperation on the Nile (ICCON), *Nile Basin Initiative: Strategic Action Program—Overview* (Prepared by the Nile Basin Initiative Secretariat in Cooperation with the World Bank, May 2001), at iii.

[237] The NBI Secretariat is headed by an Executive Secretary. The Headquarters Agreement was concluded between the Government of the Republic of Uganda and the NBI in Kampala on November 4, 2002, following the enactment by Uganda of the Nile Basin Initiative Act (*supra* n. 235). As will be discussed later, the NBI Secretariat has started recently to undertake the notification of the Nile riparians for Bank-financed projects on behalf of any of the Nile Basin countries. *See infra* n. 387.

[238] In addition to the NBI Secretariat, the organs of the NBI include the Council of Ministers of Water Affairs of the Nile Basin Countries (Nile-COM), which provides policy guidance and makes decisions on matters relating to NBI, and the Technical Advisory Committee which renders advice to the Nile-COM. *See supra* n. 235.

is to assist in establishing a collaborative environment for equitable utilization of the Nile. Such collaborative environment will decrease the likelihood of disputes, and if such disputes do erupt, the aim is to have in place a framework and mechanisms for addressing and resolving it.

In addition to these two widely known cases, there are other Bank interventions in international watercourses where the Bank played a facilitative role. This role could be a freestanding one, such as the intervention involving the conclusion of the Water Charter for the Senegal River by Mali, Mauritania and Senegal on May 18, 2002. Guinea, the fourth riparian of the Senegal River was not an original party to the Charter. The Bank was able to facilitate the engagement of Guinea which signed the Charter on March 17, 2006. Consequently, Guinea joined *Organisation pour la mise en valeur du fleuve Sénégal* (OMVS) after about 40 years of non-involvement with this organization.[239] Thus, the Bank was able to facilitate the conclusion of an inclusive legal framework for the Senegal River involving all the four riparians for the first time since 1972.

Other basins where similar efforts have been undertaken included the Guarani Aquifer shared by Brazil, Argentina, Uruguay and Paraguay. Global Environment Facility (GEF)[240] funding was provided to the four countries for jointly elaborating and implementing a common institutional and technical framework for managing and preserving the Guarani Aquifer System for current and future generations.[241]

---

[239] In fact Guinea was involved with the other three riparians of the Senegal River until the 1970s. On July 26, 1963, the four riparian states concluded the Convention Relating to the Development of the Senegal River (Revue juridique et politique (Paris) XIX year, No. 2 (April—June 1965) 299–302). However, this instrument was replaced in 1972 by the Convention Relating to the Regime of the Senegal River, (*see* FAO, *supra* n. 140, at 24–31), to which Guinea was not a party. For discussion and analysis of those and other agreements on the Senegal River, and the role of the OMVS, *see* Bonaya Adhi Godana, *Africa's Shared Water Resources—Legal and Institutional Aspects of the Nile, Niger and Senegal River Systems* (Francis Pinter Publishers 1985). For more on the OMVS *see*: http://www.omvs.org/. Less than two months after Guinea signed the Water Charter, the Bank financed the First Phase of the Senegal River Basin Multi-purpose Water Resources Development Program (P093826, 2006). The financing consisted of three credits of about US$31 million to each of Mali, Mauritania and Senegal, and a grant of about US$ 18 million to Guinea. The purpose of the financing is to assist in (i) modernizing the river basin institutions within the context of the Water Charter, (ii) improving regional water resources planning, management and development, (iii) expanding regional multi-purpose water resources infrastructure, (iv) mitigating health impacts from such infrastructure, and (v) fostering economic growth through water related sector development while improving social and environmental conditions in the Basin.

[240] For more on the GEF role and funding *see infra* n. 301.

[241] *See* Environmental Protection and Sustainable Development of the Guarani Aquifer System Project (P068121, 2002). A GEF Grant of SDR10.8 million (US$13.4 million equivalent) was extended to Argentina, Brazil, Paraguay and Uruguay for the Project.

Similar efforts took place on the Mekong River which is regulated by the 1995 Agreement on the Cooperation for the Sustainable Development of the Mekong River (Mekong, or 1995 Agreement).[242] The Agreement, *inter alia*, establishes the Mekong River Commission (MRC). Of the six riparian states of the Mekong, only Cambodia, Lao People's Democratic Republic (PDR), Thailand and Vietnam are parties to the Agreement. The other two riparians, China and Myanmar are not. However since 1996 both countries have become "dialogue partners" with the MRC.[243] This status provides them with the opportunity to participate in discussions at the MRC Council and the Joint Committee meetings. The World Bank built on that development through a GEF Project whose main objective was to assist the member states of the MRC in implementing key elements of the 1995 Agreement.[244] On April 1, 2002, the People's Republic of China signed an agreement on the provision of hydrological information on the Lancang/Mekong River.[245]

Along the same lines of cooperation, assistance to strengthen existing management capacity was provided to both, the Lake Chad Basin Commission

---

The Project included components for expansion and consolidation of the current scientific and technical knowledge about the Aquifer, joint development and implementation of Guarani Aquifer System Management Framework, based upon an agreed Strategic Program of Action, and enhancement of public and stakeholder participation. It also included a component for consideration of the potential to utilize the Aquifer's clean geothermal energy. The Grant was extended to the four countries themselves because there was no aquifer management entity similar to the Mekong River Commission *(infra* n. 244), or the Lake Chad Basin Commission *(infra* n. 246).

[242] For a copy of the Agreement *see* 34 I.L.M. 864 (1995). For more details on the Mekong River and the history of the Commission, *see* Greg Browder & Leonard Ortolano, *The Evolution of an International Water Resources Management Regime in the Mekong River Basin*, 40 Natural Resources Journal 499 (2000).

[243] *See*: http://www.mrcmekong.org/.

[244] *See* Mekong River Water Utilization Project (P045864, 2000). A GEF Grant of SDR8 million (equivalent to US$11 million) was extended to the Mekong River Commission (MRC) to assist the MRC countries in implementing key elements of the 1995 Agreement on Cooperation for Sustainable Development of the Mekong Basin. The Project's objectives included assisting the MRC to establish mechanisms to promote and improve coordinated and sustainable water management in the Basin, including reasonable and equitable water utilization by the countries of the Basin. The objectives also included protection of the environment, aquatic life and the ecological balance of the Basin. This would be achieved through preparation of "Rules" for water utilization (quantity and quality) and procedures for information exchange, and notification/consultation in accordance with the Mekong Agreement.

[245] As stated in the MRC's web site: "Under this agreement China now provides water level data in the flood season from two stations located on the Upper Mekong in China. This information is fed into the MRC's flood forecasting system. Talks are under way to expand this data sharing agreement to include dry season levels. China contributes 16% of the flow of the Mekong River." *See*: http://www.mrcmekong.org/about_mrc.htm.

(LCBC)[246] and the Niger Basin Authority (NBA).[247] Furthermore, assistance was provided to the Southern African Development Community (SADC) for a Groundwater and Drought Management Project.[248]

The Bank has also been involved with the environmental challenges facing both, the Aral Sea and the Caspian Sea. The Bank's involvement with the problems of the Aral Sea started immediately after the collapse of the Soviet Union

---

[246] *See* Reversal of Land and Water Degradation Trends in the Lake Chad Basin Ecosystem (P070252, 2003). A GEF Grant of SDR2.2 million was extended to the Lake Chad Basin Commission (LCBC) to assist in strengthening the capacity of the LCBC to better achieve its mandate of (i) coordinating LCBC's member countries' utilization of the Lake, and (ii) managing land and water resources in the greater conventional basin of Lake Chad. For the Convention and Statutes establishing LCBC, and the parties to those instruments, *see supra* n. 137.

[247] *See* Reversing Land and Water Degradation Trends in the Niger River Basin (P070256, 2004). A GEF Grant of US$6 million was extended to the Niger Basin Authority (NBA) to assist the nine riparian countries of the Basin (Benin, Burkina Faso, Cameroon, Chad, Côte d' Ivoire, Guinea, Mali, Niger and Nigeria) in defining a transboundary framework for the sustainable development of the Niger River through strengthened management capacity of the NBA, and better understanding of the Basin's land and water resources. Three years later, an IDA Credit was extended to each of Benin, Mali, and Nigeria, and an IDA Grant to each of Guinea and Niger (P093806, 2007) was extended for the Niger Basin Water Resources Development and Sustainable Ecosystems Management Program (Phase 1). The overall development objective of the Program is to enhance regional coordination, development and sustainability of water resources management in the Niger River Basin. The expected outcomes include: strengthened institutional coordination for regional management and development of water resources in the Niger River Basin, as well as improved watershed management in targeted areas in the Basin. *See* Convention Creating the Niger Basin Authority (with Protocol relating to the Development Fund of the Niger Basin) 1980, 1346 U.N.T.S. 208 (1984). For an analysis of the Convention and the previous agreements on the Niger River, *see* Salman M. A. Salman, *Niger River*, in *Max Plank Encyclopedia of Public International Law* (Oxford University Press 2009), available at: http://www. mpepil.com/subscriber_article?script=yes&id=/epil/entries/law-9780199231690-e1325&recno=1&author=Salman%20%20Salman%20MA. *See* also Inger Andersen *et al*, *The Niger River Basin—A Vision for Sustainable Management* (The World Bank 2005).

[248] *See* Groundwater and Drought Management Project (P070547, 2005). A GEF Grant of US$7 million was extended to the Southern African Development Community (SADC) to support strengthening the capacity of SADC's member states in defining drought management policies, specifically in relation to the role, availability and supply potential of groundwater resources. The primary target groups for the Project are the users of groundwater and groundwater dependent ecosystems in drought-prone areas in southern Africa. The Project would be implemented through the SADC Water Division, and the respective governments' departments water units. SADC is an inter-governmental organization headquartered in Gaborone, Botswana. The SADC member countries are: Angola, Botswana, the Democratic Republic of Congo, Lesotho, Madagascar, Malawi, Mauritius, Mozambique, Namibia, South Africa, Swaziland, Tanzania, Zambia, and Zimbabwe. Its goal is to further socio-economic cooperation and integration as well as political and security cooperation among 14 southern African states. Seychelles had previously been

and the emergence of the Central Asian Republics in the early 1990s.[249] Between 1960 and 2004, the Sea lost close to 70 percent of its surface area, and its level dropped by more than 15 meters. In 1990 the Sea split into the small Northern Aral Sea, and the larger Southern Aral Sea. As a result, serious environmental, health, social and economic problems ensued, with pollution and salinity levels rising dramatically, and fish production virtually disappearing.[250]

Working with other donors and UN agencies, the Bank was successful in focusing attention on the problem of the shrinking of the Aral Sea. The Bank was able to get the five Central Asian Republics to work together to implement a program for arresting, and gradually reversing the shrinkage, and addressing its consequences. That program included financing of projects designed to meet that objective.[251]

Although the Caspian Sea has not been facing similar problems as those of the Aral Sea, the Bank, the United Nations Environment Programme (UNEP), the United Nations Development Programme (UNDP), and the European Community (EC), together with other partners, assisted in establishing the Caspian

---

a member of SADC from September 1997 until July 2004. SADC has been established as an official international organization through the Treaty of the Southern African Development Community, April 1, 1980, 32 I.L.M. 120 (1993). *See* generally, Salman M. A. Salman, *Legal Regime for Use and Protection of International Watercourses in the Southern African Region: Evolution and Context*, 41 Natural Resources Journal, 981 (2001).

[249] The coastal lines of the Aral Sea are shared by Kazakhstan and Uzbekistan. The Amu Darya and the Syr Darya are two major rivers flowing into, and feeding the Aral Sea, and as such they are the main tributaries of the Sea. These rivers are also shared by Afghanistan, Kazakhstan, the Kyrgyz Republic, Tajikistan, Turkmenistan, and Uzbekistan.

[250] The problems of the Aral Sea resulted mainly from over-abstraction of large quantities of the waters of the Amu Darya and Syr Darya rivers for cotton production in Kazakhstan and Uzbekistan during the Soviet era, leaving a limited amount of water to reach the Aral Sea. For more details on those problems and the current Bank program in the Aral Sea *see:* http://web.worldbank.org/WBSITE/EXTERNAL/COUNTRIES/ECAEXT/EXTECAREG TOPENVIRONMENT/0,,contentMDK:20633813~menuPK:511452~pagePK:2865114~ piPK:2865167~theSitePK:511433,00.html. For discussion of the role of the Bank in the initial stages of its involvement with the Aral Sea states, *see* Kirmani & Moigne, *supra* n. 229.

[251] One of the Bank-financed projects is Kazakhstan, the Syr Darya Control and Northern Aral Sea (NAS) Phase I Project (P046045, 2001). The Project seeks to sustain and increase agriculture and fish production in the Syr Darya Basin, and secure the existence of NAS by improving ecological and environmental conditions in the delta area. Project components include restoration of the NAS area through construction of a closure dike in the Bering Strait, a channel connecting the NAS and the larger Southern Aral Sea (LAS), and conveying additional flows into the NAS. Furthermore, the global effort resulted, *inter alia,* in the establishment of the International Fund for Saving the Aral Sea (IFAS) by the heads of state of the five Central Asian Republics. The Agreement on the Status of IFAS and its Organizations was concluded by the five states on May 29, 1997. For more information on IFAS and its organizations and the Agreement, *see:* http://www.ec-ifas.org/ English_version/About_IFAS_eng.htm.

Environment Program (CEP) in the mid 1990s. The CEP aims at sustainable development of the Caspian environment, including living resources and water quality, protecting human health and ecological integrity for the sake of future generations. One of its main objectives is to assist the Caspian littoral states to achieve the goal of environmentally sustainable development and management of the Caspian environment for the sake of long-term benefit for the Caspian inhabitants.[252]

As a result of the CEP, the five littoral states concluded the Framework Convention for the Protection of the Marine Environment in the Caspian Sea (the Tehran Convention).[253] This is the first time that all the littoral states of the Caspian Sea have entered into a legally binding agreement. It should be clarified that the CEP limited itself to the environmental challenges facing the Caspian Sea, and has not addressed the issue of the territorial boundaries of each of the five littoral states within the Caspian Sea. As will be discussed in Part 9.2 of this Book, this issue is still without a resolution.

Hence, the policy has addressed the involvement of the Bank on international waterways in two parallel, but complementary tracks. The first is the facilitative efforts involving assistance to riparians for resolution of existing disputes, avoidance of future ones, and establishment of regulatory and institutional frameworks involving all the riparians. The other track involves notification of all the riparians of the Bank-financed projects. The involvement of the Bank in the facilitative efforts includes both fresh waters, as well as closed seas. This is in line with the Bank's definition of the term "international waterways." The next Parts of this Chapter will deal with the definition of international waterways, the types of projects covered by the Bank policy, and the substantive rules set forth under the policy.

4.3   Waterways Covered under the Policy

As may be recalled, the title of the 1956 OM was "Projects on International Inland Waterways." The 1965 OM changed the title to "Projects on International Waters." However, the OM itself dealt with two categories, namely (i) international inland

---

[252] The Caspian Sea is bounded by Azerbaijan, Iran, Kazakhstan, the Russian Federation and Turkmenistan. It is the largest landlocked sea in the world. For discussion of the role of the Bank and the other partners in the CEP, *see* Rie Tsutsumi & Kristy Robinson, *Environmental Impact Assessment and the Framework Convention for the Protection of the Marine Environment of the Caspian Sea,* in *Theory and Practice of Transboundary Environmental Impact Assessment,* at 53 (Kees Bastmeijer & Timo Koivurova, eds., Martinus Nijhoff Publishers 2008). A more detailed description of the Bank role in the CEP is available at: http://go.worldbank.org/03P7VUVWW0.

[253] The Convention was concluded in Tehran, Iran, in November 4, 2003. It entered into force on August 12, 2006. The Convention is available at: http://www.caspianenvironment.org/newsite/Convention-FrameworkConventionText.htm.

waterways which included any river, canal, lake or other inland waterway; and (ii) international waters which included bays, gulfs, straits, or channels bounded by several states, or if within one state recognized as necessary channels of communication between the open sea and other states. The scope of the waters covered was maintained in OM 2.22 of 1971, as well as OMS 2.32 of 1977.

The 1985 OMS reverted back to the title of the 1956 OM, but without the word "inland." The title of OMS 2.32 of 1985, and of the subsequent directives and operational policies has been "Projects on International Waterways." The term "international waterways" has since become the official Bank term, and is used throughout the policy to refer to both fresh waters and semi-enclosed coastal waters.

As discussed earlier, the IIL started using the term "watercourses" in 1911 when it issued the "Madrid Declaration."[254] At that time the connotations of the term were limited to rivers. That term was used in the arbitration award concerning the Lake Lanoux dispute in 1957 to extend to both rivers and lakes.[255] The IIL continued using the term "watercourse," and its Salzburg Resolution issued in 1961 deals "with the utilization of waters which form part of a watercourse or hydrographic basin which extends over the territory of two or more States."[256] The Resolution also uses the term "hydrographic basin" in a manner synonymous with the term "watercourse."[257]

The ILA, on the other hand, adopted a different approach. Although the first statement issued by the ILA in 1956 in Dubrovnik[258] dealt with international rivers, the ILA soon after that moved to the term "drainage basin." The New York Resolution issued in 1958 stated that "A system of rivers and lakes in a drainage basin should be treated as an integrated whole (and not piecemeal)."[259] This concept was refined and expanded to groundwater in the Helsinki Rules issued in 1966.[260] The Helsinki Rules define the term "international drainage basin" as "a geographical area extending over two or more States determined by the watershed limits of the system of waters, including surface and underground waters,

---

[254] *See supra* n. 64.

[255] *See supra* n. 82.

[256] *See* Article 1, Salzburg Resolution, *supra* n. 69.

[257] Article 1 of the Salzburg Resolution states that: "The present rules and recommendations are applicable to the utilization of waters which form part of a watercourse or hydrographic basin which extends over the territory of two or more States."

[258] *See* Dubrovnik Statement, *supra* n. 72, at 241.

[259] *See* Article 1 of the New York Resolution, *supra* n. 163.

[260] ILA, Report of the Fifty-Second Conference (Helsinki 1966), at 486. *See* also *supra* n. 166.

flowing into a common terminus."[261] However, the ILA started using the term "watercourses" synonymously with the term "drainage basin." As discussed earlier, the ILA used this term in 1976 when it issued the "Rules on Administration of International Watercourses." The term was used again in the two sets of rules adopted in 1980. The first set dealt with the regulation of the flow of the water of international watercourses, and the second dealt with the relationship of international water resources to other natural resources' environmental elements.[262]

In parallel, the United Nations General Assembly used the term "watercourse" in 1970 when it asked the ILC to study the topic of international watercourses with the view of its progressive development and codification.[263] The term "international watercourses" was used by the successive rapporteurs of the ILC, and was finally incorporated in the UN Watercourses Convention.[264] As stated earlier, the Convention was adopted by the United Nations General Assembly on May 21, 1997.[265] Article 2 of the Convention defines the term "watercourse" to mean "a system of surface waters and groundwaters constituting by virtue of their physical relationship a unitary whole and normally flowing into a common terminus."[266] Thus, the term "watercourse" has gradually gained international acceptance and is now widely used to include both surface and groundwater.[267]

---

[261] *See* Article 2 of the Helsinki Rules. It should be clarified that this definition includes only aquifers that are connected to surface water. Transboundary aquifers that do not contribute water to, or receive water from surface waters of an international drainage basin are not covered by this definition. As will be discussed later, the ILA issued the Seoul Rules on International Groundwater in 1986, which extended the definition to cover those types of aquifers. For the Seoul Rules *see* ILA, Report of the Sixty-Second Conference (Seoul 1986) at 275.

[262] *See supra* n. 171.

[263] *See supra* n. 175.

[264] For the manner in which the different rapporteurs used the term "international watercourse" and how the meaning of the term evolved, *see* Krishna, *supra* n. 130, at 37–9.

[265] *See supra* n. 180.

[266] Like the Helsinki Rules, the definition of watercourse under the Convention includes only groundwater connected to surface water, and does not extend to aquifers that do not contribute water to, or receive water from shared surface waters. The ILC, after completing the work on the UN Convention in 1994, passed in the same year a "Resolution on Confined Transboundary Groundwater" to address this lacuna, as will be discussed in Chapter 8 of this Book.

[267] The term "watercourse" is now used in a number of international instruments. The United Nations Economic Commission for Europe (UNECE) adopted in 1992 the "Convention on the Protection and Use of Transboundary Watercourses and International Lakes" (*see* 31 I.L.M. 1312 (1992)). The members of the Southern African Development Community adopted in 1995 the "Protocol on Shared Watercourse Systems in the Southern African Development Community (SADC) Region." That Protocol was revised and reissued in 2000 as the "Revised Protocol on Shared Watercourses in the Southern African Development Community (SADC)" *see supra* n. 57. The Zambezi Agreement concluded on July 13, 2004

During the revisions of the policy in 1984–85, the Bank considered the term "drainage basin" as well as "watercourse." Clearly, the term "watercourse" has acquired a specific meaning related to rivers, lakes, and groundwater. Given that the Bank policy extends to both surface waters and semi-enclosed coastal waters, the term "watercourse" would not be appropriate. Moreover, the Bank has used the term "waterways" since 1956, and the term has become a brand name of the Bank policy.[268] Accordingly, the policy kept the term "international waterways"[269] but expanded the definition considerably. The policy covers three types of waterways:

The first category of waterways includes any river, canal, lake or similar body of water that forms the boundary between, or any river or body of surface water that flows through, two or more states, whether such states are Bank members or not. This part of the definition is similar to the one introduced in 1956, and it has continued throughout the different reiterations of the policy.

The second category includes any tributary or other body of surface water that is a component of any waterway described above. This category was not included in the previous directives and was added in 1985. The addition was meant to address a basic gap in the definition of rivers and lakes under the previous directives. Tributaries or other components of shared rivers or lakes would henceforth be covered by the policy. The policy does not indicate what is meant by a tributary[270] or component, but with the extension of the policy to tributaries and other components, the entire shared basin has come under the ambit of the policy. This extension brought the definition of the term "waterway" for those two categories under the policy close to that of "watercourse" or "drainage basin" used initially by the ILA.[271]

---

also used the term "watercourse," *see* Agreement on the Establishment of the Zambezi Watercourse Commission available at: http://www.zacpro.org/downloads/ZAMCOM%20AGREEMENT.pdf.

[268] As discussed earlier, the Barcelona Convention used the term "waterway" to refer to both fresh and sea waters, *see supra* n. 55.

[269] OMS 2.32 of 1985 used the term "waterway system." The term "system" has been used for a number of river basins, such as the Indus system of rivers; *see* Preamble to the Indus Waters Treaty 1960, *see supra* Chapter 4, Part 4.2 of this Book. However, the term "system" was dropped from the subsequent versions of the Bank policy.

[270] The Indus Waters Treaty defines the term "tributary" to mean " . . . any surface channel, whether in continuous or intermittent flow and by whatever name called, whose waters in the natural course would fall into that river, e.g. a tributary, a torrent, a natural drainage, an artificial drainage, a *nadi*, a *nallah*, a *nai*, a *khad*, a *cho*. The term also includes any sub-tributary or branch or subsidiary channel, by whatever name called, whose waters, in the natural course, would directly or otherwise flow into that surface channel." *See* Article 1(2) of the Treaty, *supra* n. 219. Seasonal streams would qualify as tributaries under this definition.

[271] *See* the Helsinki Rules, *supra* n. 166.

The third category of international waterways under the policy deals with any bay, gulf, strait or channel bounded by two or more states, or if within one state recognized as a necessary channel of communication between the open sea and other states, and any river flowing into such waters. As discussed earlier, OM 5.05 added two types of bays, gulfs, straits and channels: (i) those bounded by several states,[272] such as the Gulf of Aqaba, the Gulf of Fonseca,[273] the Bay of Bengal,[274] the Strait of Malacca,[275] or the Mozambique Channel, or the Corfu Channel;[276] and (ii) those within one state but are recognized as a necessary channel of communication between the open sea and other states. The Bosporus Strait falls fully within Turkey; however, it is a necessary channel of communication between the Black Sea states, namely Bulgaria, Romania, Ukraine, Russia and Georgia, and the Mediterranean Sea.[277] Thus, the Bosporus Strait fits under this category of international waterways under the policy.

The policy does not define any of the terms "bay, gulf, strait or channel." In some instances, the Bank had to look for assistance elsewhere, mainly the 1982 UN Convention on the Law of the Sea (UNCLOS) as well as the commentaries thereon,[278] to determine whether a specific waterway described as a bay or gulf in common parlance or usage would fall under any of these categories.[279] In a number of projects the Bank made its own determination that the mere naming of

---

[272] The term "several states" should be understood to mean "two or more states" as was stated with regard to rivers and lakes in paragraph 1(b)(i) of the same OM.

[273] The Gulf of Aqaba is bounded by Egypt, Israel, Jordan and Saudi Arabia, while the Gulf of Fonseca is bounded by El Salvador, Honduras and Nicaragua.

[274] The Bay of Bengal is bounded by Bangladesh, India and Myanmar.

[275] The Strait of Malacca lies between Indonesia, Malaysia and Singapore.

[276] The Mozambique Channel falls between Mozambique and Madagascar, while the Corfu Channel is bounded by Albania and Greece.

[277] The Bosporus Strait which forms the boundary between the European part of Turkey and its Asian part is the world's narrowest strait. It is used for international navigation as it connects the Black Sea with the Sea of Marmara, which is connected by the Dardanelles to the Aegean Sea, and thereby to the Mediterranean Sea.

[278] For the United Nations Convention on the Law of the Sea (hereinafter UNCLOS) *see*: http://www.un.org/Depts/los/convention_agreements/convention_overview_conven tion.htm.

[279] Article 10 of UNCLOS defines a bay as "a well-marked indentation whose penetration is in such proportion to the width of its mouth as to contain land-locked waters and constitute more than a mere curvature of the coast. An indentation shall not, however, be regarded as a bay unless its area is as large as, or larger than, that of the semi-circle whose diameter is a line drawn across the mouth of that indentation." The other terms (gulf, strait and channel) are not defined under the UNCLOS.

a particular waterway as "bay, gulf, strait or channel" would not necessarily require the Bank to treat it as such for the purposes of its policy.[280] In the Chad-Cameroon Pipeline Project,[281] for example, the Bank deliberated whether the Gulf of Guinea is really a gulf for the purposes of its policy. It noted that the Gulf of Guinea is an open large arm of the Atlantic Ocean stretching from the western coast of Côte d'Ivoire to the Gabon estuary, and that it is bound on the south by the equator. Accordingly, the Bank concluded that the Gulf of Guinea does not fall within the concept of gulf envisaged under its policy, and for that reason notification of the littoral states was not required for that project. In reaching that conclusion, the Bank consulted and was guided by the work of a number of experts in this field. One of those experts, Sir Cecil Hurst, stated that:

> The antithesis to a bay is open sea. No one seems to have considered that waters ceased to be part of the open sea merely because they bore the name of gulf or bay. The Bay of Bengal, or the Gulf of Oman, vast stretches of waters which lack all the characteristics of a defined inlet,

---

[280] One example of the difficulties regarding the definition of those terms could be inferred from the fact that the ICJ referred to Fonseca as "the Bay or Gulf of Fonseca." *See*: http://www.icj-cij.org/docket/files/120/14077.pdf.

[281] The Chad, Petroleum Development and Pipeline Project (P044305, 2000) was approved in 2000, and consisted of an IBRD loan of US$39.5 million to the Republic of Chad, and US$53.4 million to the Republic of Cameroon to finance part of the equity of the Government of Chad in the Tchad Oil Transportation Company, S.A. (TOTCO) and the Cameroon Oil Transportation Company, S.A. (COTCO). It involved drilling of about 300 wells in Southern Chad and the construction of about 1,100 km long pipeline, mostly running through Cameroon, and extending into a subsea pipeline for approximately 11 kilometers, to an off-shore loading facility. The Project was accompanied in Chad by: (i) the IDA-financed Management of the Petroleum Economy Project (US$7.5 million), which aimed to build Chad's capacity to manage oil revenues and to use them efficiently for poverty reduction; and (ii) the IDA-financed Petroleum Sector Management Capacity Building Project (US$23.7 million), which aimed to assist the Government of Chad in carrying out its responsibilities under the Pipeline Project Environmental Assessment and to establish an effective framework for further sound private sector investment in the petroleum sector. *See* Project Appraisal Document on proposed IBRD loans in amounts of US$39.5 million to the Republic of Chad, and US$53.4 million to the Republic of Cameroon and on proposed IFC loans in amounts of US$100 million in A-loans and up to US$300 million in B-loans to the Tchad Oil Transportation Company, S.A. and Cameroon Oil Transportation Company, S.A. for a Petroleum Development and Pipeline Project. On September 9, 2008, the World Bank announced that it was unable to continue supporting the Chad-Cameroon Pipeline Project because key arrangements that had underpinned its involvement in and support for the Project were not working, notably the agreement that the Government of Chad would allocate oil revenues for poverty-reducing projects in education, health, infrastructure, rural development and improving governance. As of September 5, 2008, the Government of Chad had fully prepaid its loans for pipeline-related financing. *See*: http://go.worldbank.org/LNOXOH2W50.

have never seriously been alleged not to form part of the open sea. They stand on the same footing as the Gulf of Mexico or the Bay of Biscaye.[282]

Despite the conclusion that the Gulf of Guinea is not a gulf for the purposes of the Bank policy, the Bank took note of the fact that the floating storage and off-loading vessel is located about 60 kilometers north of the Equatorial Guinea coastline. Given this proximity, and the risks involved, the Bank decided that Equatorial Guinea should be notified of the project. Notification was undertaken by the Bank on behalf of Chad.[283]

The determination that the Gulf of Guinea does not qualify as a gulf under the Bank policy was reconfirmed in the subsequent West African Gas Pipeline Project.[284] Similarly, the Bank considered the status of the Bay of Bengal, but since the project was for rehabilitation of existing schemes, the issue of notification of the littoral states of the Bay of Bengal was not pursued.[285] As will be discussed in Chapter 7, Part 7.2.1, projects for rehabilitation of existing schemes are dealt with under one of the exceptions to the notification requirement.

In addition to these two types of waterways, the policy added a third component to this category, namely any river flowing into such waters. The river referred to in this paragraph is a national and not international river, because

---

[282] *See* Sir Cecil Hurst, *The Territoriality of Bays*, 3 Brit. Yb. Int'l L. 42 (1922–23).

[283] The pipeline crosses several streams and rivers which flow into Lake Chad, including the Mbere River, which is a boundary river between Cameroon and Chad, and the Logone River which joins the Chari River, which flows into Lake Chad. The riparians of those waterways (the Central African Republic, Niger and Nigeria), in addition to the Lake Chad Basin Commission, were notified of the project. The Central African Republic asked for more information, which was supplied, and then it gave its no-objection to the Project. Niger, like Equatorial Guinea, also gave its no-objection. Notification to Equatorial Guinea was prompted by the environmental impact assessment prepared under Bank OP 4.01 which highlighted the proximity of the pipeline and the risks to Equatorial Guinea.

[284] The Project consisted of the construction of a pipeline system that would transport natural gas from Nigeria to Ghana, Togo and Benin. The Project included spurs to provide gas-to-power generating units in Ghana, Benin, and Togo, the conversion of existing power-generating units to gas, and, as needed, additional compression investments. The new pipeline (678 kilometers long) originates at a connection to an existing pipeline in Nigeria. Fifty-eight kilometers of pipeline and other ancillary facilities are to be constructed in southwestern Nigeria, and the pipeline then runs off-shore to a terminal point in Takoradi, Ghana. *See* 3A-West African Gas Pipeline (IDA S/UP) (P082502, 2004).

[285] The India Coal Fired Generation Rehabilitation Project (P100101, 2009) would renovate and modernize four generation units of old coal-fired generation units to test and demonstrate energy efficient rehabilitation approaches. Water for cooling the units would come from the Hooghly River, which bifurcates from the Ganges River and flows towards Calcutta and empties into the Bay of Bengal. It was concluded that the water used and change of temperature in the Hooghly River waters, as a result of the rehabilitation works, would be minimal and would have no effects on the river or the Bay of Bengal.

international rivers are covered in the first group of international waterways under the policy. This means that any national river flowing into a bay, gulf, strait or channel described above would, for the purpose of the Bank policy, be considered an international river.

No doubt, this is a major expansion of the list of international waterways under the policy. As discussed before, concerns about pollution to the bays, gulf, straits and channels as a result of Bank-financed projects, and the overall objective of the Bank policy of dispute avoidance, prompted the Bank to add semi-enclosed water bodies to its policy in 1965. The rationale for that action was the need to protect the marine environment from pollution that may be caused by Bank-financed projects.[286]

The same rationale applies to the addition of national rivers flowing into those bodies of semi-enclosed waters. This is because the most significant cause of marine pollution is from land-based sources, and this pollution is principally carried into the sea by rivers.[287] The carriers of such pollution to the coastal waters could be international, as well as national rivers. Accordingly, the flow of national rivers could have transboundary effects when the marine environment is shared by two or more states, as often is the case, thus giving the national rivers an international dimension.

Support for this international dimension of national rivers can actually be found in the UNCLOS itself. Article 207 of the Convention requires the parties to ". . . adopt laws and regulations to prevent, reduce and control pollution of the marine environment from land-based sources, including rivers, estuaries, pipelines and outfall structures, taking into account internationally agreed rules, standards and recommended practices and procedures."[288] The Article goes on to require the parties to take other measures as may be necessary to prevent, reduce and control such pollution.[289] Similar provisions are included in the 1992 Protocol on Protection of the Black Sea Marine Environment against Pollution from Land-based

---

[286] Article 1 of UNCLOS defines "pollution of the marine environment" to mean "the introduction by man, directly or indirectly, of substances or energy into the marine environment, including estuaries, which results or is likely to result in such deleterious effects as harm to living resources and marine life, hazards to human health, hindrance to marine activities, including fishing and other legitimate uses of the sea, impairment of quality for use of sea water and reduction of amenities."

[287] *See* David Freestone & Salman M. A. Salman, *Ocean and Freshwater Resources*, in *The Oxford Handbook of International Environmental Law* 337 *et seq.* (Daniel Bodansky, Jutta Brunnee & Ellen Hey, eds., Oxford University Press 2007).

[288] *See* also Catherine Redgwell, *From Permission to Prohibition: The 1982 Convention on the Law of the Sea and Protection of the Marine Environment*, in David Freestone, Richard Barnes & David M. Ong, *The Law of the Sea—Progress and Prospects* 180 *et seq.* (Oxford University Press 2006).

[289] Article 207 of UNCLOS goes on to require the state parties ". . . acting especially through competent international organizations or diplomatic conference, (to) endeavour

Sources.[290] The Protocol calls on the parties to take all necessary measures to prevent, reduce and control pollution of the marine environment of the Black Sea caused by discharges from land-based sources on their territories such as rivers, canals, and coastal establishments. Interestingly, the Protocol calls those rivers and canals "tributaries to the Black Sea" and asks the parties to cooperate to achieve the purposes of the Protocol.[291] In this connection, the Bank policy has consistently considered the Black Sea itself an international waterway, and the Commission for the Protection of the Black Sea has been notified of Bank-financed projects that may affect the Black Sea.[292] Thus, the Bank policy has effectively, and practically, considered national rivers as tributaries of both closed seas and semi-enclosed coastal waters.

One of the questions that arose in connection with the interpretation and application of the policy concerns closed seas, such as the Aral, Caspian, and Dead seas, whose coastlines, as discussed before, are shared by a number of states.[293] Are they covered by the policy, and if so, under which category? The rationale for inclusion of the semi-enclosed coastal waters, namely concerns about pollution and ecological integrity, applies equally and perhaps more strongly, to these seas as well. There was no disagreement that these seas are indeed international waterways. The question that arose is under what category of the policy are they covered? An interpretation was rendered that they fall under the first category of "similar body of water that forms a boundary. . . ." In this way, the closed seas are considered, for the purposes of the Bank policy, a body of water similar to lakes.[294]

---

to establish global and regional rules, standards and recommended practices and procedures to prevent, reduce and control pollution of the marine environment from land-based sources, taking into account characteristic regional features, the economic capacity of developing States and their need for economic development. Such rules, standards and recommended practices and procedures shall be re-examined from time to time as necessary."

[290] The Protocol was concluded in connection with the 1992 Convention on the Protection of the Black Sea Against Pollution (The Bucharest Convention), *see* 1764 U.N.T.S. 4, and 32 I.L.M. 1101 (1993) for the Convention, and 1764 U.N.T.S. 18 (1994) for the Protocol.

[291] *See* Article 4 of the Protocol, which reads "As to watercourses that are tributaries to the Black Sea, the Contracting Parties will endeavour to cooperate, as appropriate, with other States in order to achieve the purposes set forth in this Article." It is worth adding that the Protocol does not distinguish between international and national rivers.

[292] The Commission is sometimes referred to as the "Istanbul Commission," because its Permanent Secretariat is located in Istanbul, Turkey. One example of notification to the Istanbul Commission was for the Georgia, Infrastructure Pre-Investment Facility (P098850, 2006).

[293] For more information on the Aral Sea *see supra* n. 249. For the Caspian Sea *see supra* n. 252, and for the Dead Sea *see infra* nn. 576 and 577.

[294] *See* discussion of the Azerbaijan, Petroleum Technical Assistance Project (P008282, 1995) where the Bank confirmed that the Caspian Sea is indeed covered by the policy,

Accordingly, national rivers flowing into the Aral or Caspian sea would also be covered by the definition of international waterways under the policy, as tributaries of those bodies of water. The Volga River which originates and flows entirely within the Russian Federation empties into the Caspian Sea, and is thus an international waterway for the purposes of the policy because it is a tributary of the Caspian Sea.

However, when these rivers flow into closed seas, the concerns extend to water quantity issues and not just pollution, as is the case with semi-enclosed coastal waters. A significant and continuous decrease in the flow of any national river into any of these closed seas could adversely affect the ecological integrity of such a sea. The Aral Sea, as discussed before, has shrunk considerably as a result of a decrease in the flow of the Amu Darya and Syr Darya rivers, and the Dead Sea as a result of the decrease of the flow of the Jordan River. In Bank-financed projects on national rivers flowing into such closed seas, the Bank ensures that that both the quality and quantity of water flow is not affected by such projects, and that the littoral states of such closed seas are notified of such projects.

As mentioned earlier, when the first part of this category of international waters was added to the Bank policy in 1965, the Bank did not have an operational memorandum on the environment. However, in May 1984, Operational Manual Statement (OMS) 2.36 on the Environment Aspects of Bank Work was issued outlining the Bank policies and procedures for projects that may have environmental implications.[295] The OMS stated that the Bank would not finance projects that could significantly harm the environment of a neighboring country without the consent of that country.[296] These types of projects which could affect semi-enclosed waters were already covered by the 1965. However, one year after OMS 2.36 was issued, the Bank, through its OMS 2.32, maintained and expanded the list of semi-enclosed coastal waters basically to underscore the concerns about pollution. One reason for maintaining and expanding the list of international waterways under the Bank policy has been the recognition by the Bank of the need to notify and assure the other littoral states that Bank-financed projects on these waterways would not cause any adverse effects. OMS 2.36 did not include any provisions on notification, and for that reason the reference to those coastal waters has been kept in the policy for projects on international waterways.

The definition of international waterways under the policy does not cover any type of transboundary groundwater. As will be discussed later, by the time

---

infra n. 680. Similarly, *see* also discussion of the Kazakhstan, Second Irrigation and Drainage Improvement Project (P086592, 2009), *infra* n. 538.

[295] *See supra* n. 109.

[296] *See* paragraph 9(f) of OMS 2.36.

the 1985 OMS was issued, only the 1966 Helsinki Rules had made an explicit reference to transboundary groundwater. This reference was further elaborated and expanded by the Seoul Rules issued by the ILA in 1986, one year after the OMS was issued.[297] As may be recalled, in 1984 and 1985, when OMS 2.32 was being prepared, the ILC was still in the early stages of its work on the draft convention, and had not, by that time, agreed on a definition for the term "international watercourses." Because of that, as well as the technical difficulties surrounding transboundary groundwater in the mid-1980s, the Bank was reluctant to include transboundary groundwater in the scope of waters covered by the policy. However, as will be discussed later, since 1990, the Bank has extended application of its policy for projects on international waterways to shared aquifers, and has required notification of the states that share such an aquifer of any project affecting such shared aquifer.[298]

Thus, the definition of international waterways under the Bank policy has the peculiar elements of inclusion of semi-enclosed coastal waters, closed seas, and national rivers flowing into those waters, and the absence of explicit provisions on transboundary groundwater. This definition of international waterways under the policy is at variance with the definitions under the IIL and the ILA rules, as well as the UN Watercourses Convention. However, this unique characteristic of the Bank policy can be attributed to, and indeed understood, in the historical and practical contexts of the policy, and the character of the Bank as an international financial cooperative institution.

4.4　Projects Covered under the Policy

As discussed earlier, OM 8 of 1956 listed certain types of projects covered under the Bank policy. The list included hydroelectric, irrigation, flood control, navigation, drainage, or similar projects. The 1965 OM added sewage and industrial projects, and the 1985 OMS added water projects. Although this list is quite comprehensive, the policy still includes the all-encompassing phrase ". . . and similar projects that involve the use or potential pollution of international waterways. . . ."[299] This is a fairly exhaustive list of projects. Indeed, any physical activity financed by the

---

[297] *See* Seoul Rules, International Law Association, Report of the Sixty-Second Conference, Seoul, 1986, at 238. For more detailed discussion of this issue, *see* also *infra*, Chapter 8, Part 8.2 of this Book.

[298] *See infra*, Chapter 8, Part 8.4 of this Book.

[299] The phrase "or similar project" was introduced in the 1956 OM. The phrase was expanded in the 1965 OM to read "or similar project that involves use or pollution of international waterways." The 1985 OMS reiterated that encompassing phrase.

Bank on an international waterway that could affect the quantity or quality of water would most likely fall under this list, and trigger the policy.

Still, the policy has gone beyond this exhaustive list, and included detailed design and engineering studies for the projects specified in this list. Those projects are covered under the policy regardless of whether the Bank finances them or is acting as an executing agency or in any other capacity. The expansion of the list of projects to include detailed design and engineering studies stems from two factors:

First, the obligation of the Bank to act prudently in the interests of all its members, together with its guiding principle of dispute avoidance, both dictate that the Bank take extra care when handling any kind of activities on international waterways. Detailed design and engineering studies for projects on international waterways pave the way for carrying out those projects, whether through Bank funding, financing by other donors, or from the country's own resources. Accordingly, such activities need to fall under the ambit of the policy at that stage, and not when they mature into projects. The Bank, as an international financial cooperative institution, cannot overlook other riparians' interests and rights even at that early stage.

Second, the Bank has been acting as an executing agency for a large number of trust funds contributed by a large group of donors.[300] Since 1991, the Bank has been one of the implementing agencies for the Global Environment Facility (GEF) projects.[301] International waters occupy a considerable part of the GEF program. GEF projects on international waters include both investment as

---

[300] As of December 31, 2008, the World Bank Group (WBG) held a total of $26.2 billion in trust ($18.2 billion in cash and investments, and $8 billion in promissory notes). These resources are being administered through 1,015 active funds supported by 339 sovereign and non-sovereign donor agencies. For more details on the trust funds administered by the WBG, *see*: http://web.worldbank.org/WBSITE/EXTERNAL/EXTABOUTUS/ORGANIZATION/CFPEXT/0,,contentMDK:20135627~menuPK:64060203~pagePK:64060249~piPK:64060294~theSitePK:299948,00.html.

[301] The Global Environment Facility (GEF) is a global partnership among (as of January 2009) 178 countries, international institutions, NGOs, and the private sector to address global environmental issues. The GEF is the designated financial mechanism for a number of multilateral environmental agreements (MEAs) or conventions, and in that capacity the GEF assists countries in meeting their obligations under any of those conventions to which they are parties. These conventions are the Convention on Biological Diversity (CBD); the UN Framework Convention on Climate Change (UNFCCC); the Stockholm Convention on Persistent Organic Pollutants (POPs); and the UN Convention to Combat Desertification (UNCCD). The GEF also supports projects and programs on international waters despite the lack of a convention in force in this area. Thus, the GEF provides grants for projects related to the following six focal areas: biodiversity, climate change, international waters, land degradation, the ozone layer, and persistent organic pollutants. For more information on the GEF, *see*: http://www.gefweb.org/interior.aspx?id=50.

well as technical assistance projects.[302] The Bank's operational policies, including the policy for projects on international waterways, have been applied to such activities when the Bank is acting as an implementing agency of the GEF.[303]

Detailed design and engineering studies need to be distinguished from water resources surveys and feasibility studies. For the former group of activities the policy applies, and notification is required. With regards to water resources surveys and feasibility studies, although the policy applies, those activities fall under the exception to the notification requirement, as will be discussed later. Thus, the challenge would be on how to distinguish between the two categories of activities. The mere calling of an activity a feasibility study is not sufficient to exempt it from notification, and an analysis of the detailed terms of reference would need to be undertaken to ascertain which category of activities it would fall under. There would still be a gray area falling between the two, for which judgment will need to be exercised.

The listing of the projects and studies covered by the policy is another area where the policy has approached the matter differently from the IIL and ILA rules, and the Watercourses Convention. Those instruments use an all encompassing term rather than list the different types of projects covered thereunder. The IIL Salzburg Resolution covers "works or utilization of the waters of a watercourse or hydrographic basin."[304] The ILA Helsinki Rules deal with "construction or installations,"[305] while the UN Watercourses Convention devotes Part III to "planned measures."[306] As we have seen, the list of the projects covered by the Bank policy evolved gradually between 1956 and 1985. Since the list is quite inclusive and covers almost all types of projects that the Bank finances, the Bank has kept that list. This approach has clearly been preferred to an all-encompassing term, such as works, planned measures, or projects. In this way, the policy has provided clarity and ensured consistency by specifying the projects it covers.

Despite this exhaustive list of projects, the Bank still faces questions as to whether certain types of projects are covered by the policy. One of the early

---

[302] By January 2009, 134 projects on international waters had been financed by the GEF, with funding of close to one billion dollars. For more information on those projects, *see*: http://www.gefonline.org/projectList.cfm?focalSearch=I.

[303] Other organizations that are acting as implementing agencies of the GEF include the UNDP and UNEP. For the list of these organizations, *see*: http://www.gefweb.org/interior.aspx?id=104.

[304] *See* paragraph 4 of the Salzburg Resolution, *supra* n. 69.

[305] *See* paragraph XXIX(2) of the Helsinki Rules, *supra* n. 166.

[306] *See* Article 12 of the UN Watercourses Convention, *supra* n. 180.

questions that arose in this connection related to oil and gas pipelines projects. Are such projects covered because of the likely effects on an international waterway, and if so, under which category of projects do such projects fall? The phrase under paragraph 2(a) of the policy that refers to ". . . similar projects that involve the use or potential pollution of international waterway" provided an all-encompassing umbrella for projects that are not specifically mentioned in the policy, but could have effects on international waterways. Accordingly, oil and gas pipeline projects would be covered by the policy because there is the potential for pollution of international waterways. For this reason, and as discussed earlier, the riparian states of the international waterways which the pipeline under the Chad-Cameroon Petroleum Development and Pipeline Project crosses, were notified of the project.[307]

Bridges on international waterways posed the question of whether they could involve the potential pollution of an international waterway, and thus should be covered by the policy. The World Bank financed the Jamuna Bridge Project in Bangladesh which was constructed across the Brahmaputra River.[308] The River originates in China and flows through India and Bhutan to Bangladesh where it is known as the Jamuna River, before joining the Ganges River and flowing into the Bay of Bengal. With a length of about five kilometers, the bridge is one of the longest in the world. There was no discussion of the applicability of OP 7.50 to this project despite the fact that the bridge was built across an international river. Perhaps one reason for lack of deliberation on the applicability of the policy is the fact that bridges are not specifically included in the list of projects covered under the policy. Another reason could be that if there were any impacts, they would relate to quality issues, and that Bangladesh is the lowest downstream riparian of the Brahmaputra/Jamuna River.[309]

In some instances the projects may go beyond being physical activities on an international waterway, and may entail extraction of resources from the waterways themselves. Such projects would involve the use of the shared waters to reach such resources, and could be treated as industrial projects. Fisheries projects have been

---

[307] *See supra* n. 281.

[308] *See* Bangladesh, Jamuna Bridge Project (P009509, 1994), Staff Appraisal Report (SAR) No. 12404-BD, dated January 24, 1994.

[309] The SAR indicated that, although the Project was unlikely to have significant adverse impacts on the region, a number of activities were supported under the Project, including: (i) monitoring back water effects and impacts upon agriculture; and (ii) monitoring of groundwater levels through existing automatic water level recording wells in the region. *See id.*, at 82. In connection with Bangladesh being the lowest downstream riparian of the Brahmaputra/Jamuna River, it should be remembered that the River flows into the Bay of Bengal.

dealt with as falling under this category, particularly when the project involves other activities on the shared waterway,[310] or when the fisheries project is large in scale.

Another example relates to the extraction of methane gas from a shared waterway. Lake Kivu is shared by Rwanda and the Democratic Republic of Congo (DRC), and its depths reportedly contain huge quantities of dissolved gas.[311] Precedents under the policy have established the rule that any project for the extraction of gas from an international waterway would fall under the category of industrial projects, or under the overall umbrella of "similar projects." In this particular case, it was decided that the methane is a shared resource and should only be extracted through an agreement between the two countries, and not through notification. In this way, the Bank would avoid creating a dispute, or exacerbating existing ones, and would be acting in the interests of all parties.[312]

One of the questions that the Bank has faced relates to projects that include a component on an international waterway, which the Bank is not financing, but is limiting its funding to other components of the same project. The funding for the component on the international waterway could come from another financier, or from the borrower's own resources. The Bank has long determined that under those types of projects, the policy for projects on international waterways still applies notwithstanding the fact that the Bank is not financing the component on the international waterway. This determination has been made to avoid the perception that the component to which the policy applied was intentionally carved out for other financiers simply to avoid application of the policy. Indeed, OP 4.01 on Environmental Assessment reconfirms this principle, stating that the policy applies to all components of the project, regardless of the source of financing.[313] This approach of OP 4.01 has been applied to all projects, including projects on international waterways.

---

[310] The Albania, Pilot Fishery Development Project (P069479, 2002) affects both Lake Shkorda (shared by Albania and Serbia), and Lake Ohrid (which Albania shares with the former Yugoslav Republic of Macedonia). The Project included landing sites for fishing boats, buildings and equipment. The riparians were notified of the Project.

[311] It is reported that Lake Kivu contains about 55 billion cubic meters (72 billion cubic yards) of dissolved methane gas at a depth of 300 meters. For more information on the proposed project, *see*: http://www-wds.worldbank.org/external/default/main?pagePK= 64193027&piPK=64187937&theSitePK=523679&menuPK=64187510&searchMenu PK=64187283&theSitePK=523679&entityID=000104615_20061219094806&search MenuPK=64187283&theSitePK=523679.

[312] Processing of the project was discontinued for reasons not related to OP 7.50. However, recent reports indicated that Rwanda has signed agreements for the extraction of the methane, *see*: http://news.prnewswire.com/DisplayReleaseContent.aspx?ACCT=ind_ focus.story&STORY=/www/story/03-02-2009/0004981196&EDATE.

[313] *See* n. 1 of OP 4.01.

A more complex matter relates to what is generally referred to as "connected projects." The issue here does not relate to one of the components of the same project. The issue relates to different, but somehow connected projects. One example concerns a pipeline to be financed by the Bank that runs over land, but connects to another pipeline already under construction and about to be completed, not financed by the Bank, which runs across an international waterway. Should the riparians of the waterway be notified although the Bank project only runs through land? Similarly, what if the Bank is financing a water treatment plant project on one international waterway, and the borrower is financing separately another water treatment plant which is not part of the project on another international waterway, and both plants would use the water distribution network financed under the project? Should notification be limited to the riparians of the waterway under the first treatment plant, or should it include the riparians of the waterways of both treatment plants?

Those are difficult questions and the Bank policy for projects on international waterways does not provide any guidance on how to deal with them. In those types of situations, the Bank would most likely apply, as a general rule, the criteria of "significant and direct relationship" between the two projects. If such a significant and direct relationship exists, then the policy might apply. The elements of the criteria for this relationship are now set forth in more details in the Bank Policy on Involuntary Resettlement.[314] The criteria include (i) direct and significant relationship between the two activities, (ii) necessity of the other activity for achieving the objectives of the Bank-financed project, and (iii) contemporaneous implementation of the other project with the Bank-financed project.[315] Applying these criteria, both of the above examples would require notification of the affected riparians, even though the other project is not financed by the Bank. The overall objectives of the Bank policy of dispute avoidance and of acting prudently in the interests of all members would require application of the policy. Thus, the leveraging effect of the policy, in cases of both, different components and different projects, is clearly quite substantial.

---

[314] OP/BP 4.12, issued in December 2001.

[315] Paragraph 4 of the Bank Policy on Involuntary Resettlement deals with "Impacts Covered." It states that: "This policy applies to all components of the project that result in involuntary resettlement, regardless of the source of financing. It also applies to other activities resulting in involuntary resettlement, that in the judgment of the Bank, are: (a) directly and significantly related to the Bank-assisted project, (b) necessary to achieve its objectives as set forth in the project documents; and (c) carried out, or planned to be carried out, contemporaneously with the project." The Policy is available at: http://go.worldbank. org/WTA1ODE7T0.

It should be clarified, however, that the policy for projects on international waterways does not apply to development policy lending. This type of lending supports policy reforms, as opposed to investment projects.[316] Thus, application of the policy is limited to investment lending involving certain types of projects specified in the policy, as well as detailed design and engineering studies for such projects.

In addition to the lists of waterways and projects, the policy has addressed both, the substantive principles and procedural rules regarding the financing of projects on international waterways. The next Part of this Chapter deals with the substantive principles, while the procedural rules will be addressed in the next Chapter.

4.5   Substantive Rules of the Policy

As discussed earlier, the work of the IIL has dealt mainly with the obligation of each watercourse state not to cause harm to the other riparian states, while utilizing the shared watercourse, whereas the ILA work emphasized the concept of equitable and reasonable utilization of the shared watercourse. The approach of the IIL is clearly manifested in the three earlier resolutions it issued, namely, the Madrid Declaration of 1911, the Salzburg Resolution of 1961, and the Athens Resolution of 1979. The three resolutions that the IIL adopted in 1997 at its meeting in Salzburg dealt with the environment, and are entitled "the Environment," "Responsibility and Liability for Environmental Damage under International Law," and "Procedures."[317] In line with its previous rules, these three Resolutions also emphasize the obligation of the riparian states not to cause unlawful harm to other riparians.

On the other hand, the work of the ILA, particularly the Helsinki Rules, adopted and elaborated the concept of equitable and reasonable utilization. As discussed earlier, the Helsinki Rules include a number of factors to be used for determining the equitable and reasonable utilization of each riparian state. These factors have been widely accepted, and by and large, incorporated by the ILC in Article 6 of the UN Watercourses Convention.[318] That Article, together

---

[316] Development policy lending (DPL) is dealt with under OP 8.60 issued in 2004. This OP provides a uniform framework for all Bank lending that supports policy reforms. In particular, it unifies the policy applying to a whole range of instruments, including sectoral adjustment loans, structural adjustment loans and poverty reduction support credits, under the heading of "development policy lending." For more of DPL and a copy of OP 8.60 *see*: http://web.worldbank.org/WBSITE/EXTERNAL/TOPICS/ENVIRONMENT/EXTEEI/0,,contentMDK:20487782~menuPK:2853302~pagePK:210058~piPK:210062~theSitePK:408050,00.html.

[317] *See* 67 Annuaire de l'Institut de Droit International 217 (1997).

[318] For a comparison of the factors under the Helsinki Rules and those under the UN Watercourses Convention, *see* Salman & Uprety, *supra* n. 61, at 28–29. *See also* Salman

with Article 5 of the Convention, both deal with the concept of equitable and reasonable utilization.

The Convention also deals in Article 7 with the obligation not to cause significant harm, and requires the watercourse states to take all appropriate measures to prevent the causing of significant harm to other watercourse states. Agreement on which of the two principles (equitable and reasonable utilization, or the obligation not to cause harm) takes priority over the other proved quite difficult, and the issue occupied the ILC throughout its work on the Convention. Each rapporteur dealt with the issue differently, either equating the two principles, or subordinating one principle to the other. The issue was discussed in the Working Group[319] where sharp differences between the riparian states on those two principles dominated the discussion. It is worth clarifying in this connection that lower riparians tend to favor the "no harm" rule, as it protects existing uses against impacts resulting from activities undertaken by upstream states. Conversely, upper riparians tend to favor the principle of equitable and reasonable utilization, because it provides more scope for states to utilize their share of the watercourse for activities that may impact on downstream states. After a lengthy debate by the Working Group, a compromise regarding the relationship between the two principles was reached. The compromise addressed and linked Articles 5 and 6 (equitable and reasonable utilization) and Article 7 (obligation not to cause significant harm). The new language of Article 7 requires the state that causes significant harm to take measures to eliminate or mitigate such harm "having due regard to Articles 5 and 6." As stated above, these two articles deal with the principles of equitable and reasonable utilization.[320]

However, notwithstanding this compromise language, most experts in the field of international law believe that the Watercourses Convention has subordinated the obligation not to cause significant harm to the principle of equitable and reasonable utilization.[321] This conclusion is based on a close reading of Articles 5, 6 and 7 of

---

M. A. Salman, *The Helsinki Rules, the United Nations Watercourses Convention and the Berlin Rules, supra* n. 172.

[319] As discussed earlier, the Sixth Committee of the UN (the Legal Committee) was convened as a Working Group of the Whole; *see supra* n. 177.

[320] For an elaboration of how the Working Group reached this compromise *see* Lucius Caflisch, *supra* n. 56, at 13–15. Note, in particular, Caflisch's statement that "The new formula was considered by a number of lower riparians to be sufficiently neutral not to suggest a subordination of the no-harm rule to the principle of equitable and reasonable utilization. A number of upper riparians thought just the contrary, namely that, that formula was strong enough to support the idea of such a subordination." *See id.*, at 15.

[321] *See* McCaffrey, *supra* n. 185, and Richard Paisley, *Adversaries into Partners: International Water Law and the Equitable Sharing of Downstream Benefits*, 13 Melbourne Journal of International Law 280 (2002). *See* also Freestone & Salman, *supra* n. 287, at 352.

the Convention. Article 6 enumerates a number of factors for determining equitable and reasonable utilization. Those factors include (i) "the effects of the use or uses of the watercourse in one watercourse State on other watercourse States," and (ii) "existing and potential uses of the watercourse." Those same factors will also need to be used, together with other factors, to determine whether significant harm is caused to another riparian, because harm can be caused by depriving other riparians of the water flow and thereby affect their existing uses.

Moreover, Article 7(1) of the Watercourses Convention obliges watercourse states, when utilizing an international watercourse in their territory, to take all appropriate measures to prevent the causing of significant harm to other watercourse states. When significant harm nevertheless is caused to another watercourse state, then Article 7(2) of the Convention requires the state causing the harm to "take all appropriate measures, having due regard to Articles 5 and 6, in consultation with the affected State, to eliminate or mitigate such harm, and where appropriate, to discuss the question of compensation." As noted before, Articles 5 and 6 of the Convention deal with equitable and reasonable utilization. As such, Article 7(2) requires giving due regard to the principle of equitable and reasonable utilization when significant harm has nevertheless been caused to another watercourse state. The paragraph also indicates that the causing of harm may be tolerated in certain cases such as when the possibility of compensation may be considered. Accordingly, a careful reading of Articles 5, 6 and 7 of the Convention should lead to the conclusion that the obligation not to cause harm has indeed been subordinated to the principle of equitable and reasonable utilization. Hence, it can be concluded that, similar to the Helsinki Rules, the principle of equitable and reasonable utilization is the fundamental and guiding principle of the UN Watercourses Convention, and consequently, of international water law.[322]

Almost 20 years before adopting the Watercourses Convention, the UN General Assembly had underscored the principle of the equality of states in sharing natural resources, as well as the primacy of equitable and reasonable utilization over the obligation not to cause harm. In 1979 the General Assembly issued the "Principles of Conduct in the field of the Environment for the Guidance of States in the Conservation and Harmonious Utilisation of Natural Resources Shared by Two or More States."[323] Principle 1 states that:

> It is necessary for states to co-operate in the field of the environment concerning conservation and harmonious utilisation of natural resources shared by two or more states. Accordingly, it is necessary that consistent

---

[322] As will be discussed later, *infra* n. 330, the ICJ has also endorsed this conclusion.

[323] These Principles were drafted, in response to UNGA Resolution 3129 (XIVIII) of December 13, 1973, by a UNEP working group of legal experts, which met between 1976

with the concept of equitable utilisation of shared natural resources, states cooperate with a view to controlling, preventing, reducing, or eliminating adverse environmental effects which may result from the utilisation of such resources. Such co-operation is to take place on an equal footing and taking into account the sovereignty, rights and interests of the states concerned.[324]

Along the same lines, Charles Bourne opined "Today, however, the doctrine of prior appropriation has almost been universally rejected in favour of the doctrine of equitable utilization. Under the latter doctrine, a state is always entitled to 'a reasonable and equitable share in the beneficial uses of the waters of an international drainage basin,' and 'the past utilizations of the waters of the basin, including in particular existing utilization' is only one of the many other factors to be taken into account in the determination of its share."[325]

This view has been endorsed by the International Court of Justice (ICJ) in the Danube case between Hungary and Slovakia (the *Gabčikovo-Nagymaros* case).[326] The case was decided in September 1997, four months after the Convention was adopted by the UN General Assembly. This is the first international water dispute to be referred to, and decided by, the ICJ. The dispute involves complex legal issues, including the law of treaties, state responsibility, environmental law, and the concept of sustainable development, as well as international watercourses.

The dispute arose between Hungary and Czechoslovakia regarding two barrages over the Danube River envisaged under a Treaty concluded in 1977 by the two countries. Construction began in the late 1970s, but in the mid-1980s, environmental groups in Hungary claimed negative environmental impacts of

---

and 1978. In the light of the Working Group's Report (LJNEP.IG.12/2) and further government comments on the draft principles (UN document A/34/557 and Corr.1). The General Assembly by Resolution 34/186 of 18 December 1979 requested all states "to use the principles as guidelines and recommendations in the formulation of bilateral or multilateral conventions regarding natural resources shared by two or more States, on the basis of the principles of good faith and in the spirit of good neighbourliness and in such a way as to enhance and not adversely affect development and the interests of all countries, in particular of the developing countries."

[324] This Principle goes beyond Principle 21 of the Stockholm Declaration of 1972 (Declaration of the United Nations Conference on the Human Environment) because it addresses both, equitable utilization as well the obligation to control or eliminate adverse environmental effects. Principle 21 reads "States have, in accordance with the Charter of the United Nations and the principles of international law, the sovereign right to exploit their own resources pursuant to their own environmental policies, and the responsibility to ensure that activities within their jurisdiction or control do not cause damage to the environment of other States or of areas beyond the limits of national jurisdiction."

[325] *See* Charles Bourne, *International Water Law, supra* n. 71, at 158.

[326] *See Case Concerning the Gabčikovo-Nagymaros Project* (Hungary v. Slovakia), Judgment, I.C.J. Reports 1997, General List No. 92, reprinted 37 I.L.M. 162 (1998).

the barrages and began protesting against the project, forcing the Hungarian government to suspend work in 1989. Czechoslovakia insisted that there were no negative environmental impacts, and decided to proceed unilaterally with a provisional solution consisting of a single barrage on its side, but requiring diversion of a considerable amount of the waters of the Danube to its territory. Czechoslovakia claimed that this was justified under the 1977 Treaty. As a result of the unilateral action of Czechoslovakia, Hungary decided to terminate the 1977 Treaty based on ecological necessity. The situation became more complicated with the split of Czechoslovakia in December 1992 into two countries (the Czech Republic and the Slovak Republic, or Slovakia), and the agreement that Slovakia would succeed in owning the Czechoslovakian part of the project. By that time Slovakia had already dammed the Danube and diverted the waters into its territory. The two countries agreed in April 1993, basically under the pressure from the European Commission (EC), to refer the dispute to the ICJ.

In a brief summary, the ICJ ruled in September 1997[327] that Hungary was not entitled to suspend or terminate the work on the project in 1989 on environmental grounds, and that Czechoslovakia, and later Slovakia, was also not entitled to operate the project based on the unilateral solution it developed without an agreement with Hungary. The ICJ further decided that Hungary was not entitled to terminate the 1977 Treaty on the grounds of ecological necessity,[328] and thus, the Court ruled that the Treaty was still in force. The ICJ concluded that Hungary and Slovakia must negotiate in good faith in the light of the prevailing situation, and must take all necessary measures to ensure the achievement of the objectives of the Treaty of 1977, in accordance with such modalities as they may agree upon.

The ICJ emphasized the concept of equitable and reasonable utilization when it directed that "the multi-purpose programme, in the form of a co-ordinated single unit, for the use, development and protection of the watercourse is implemented in an equitable and reasonable manner."[329] The ICJ went further in that direction, stating that it ". . . considers that Czechoslovakia, by unilaterally assuming control of a shared resource, and thereby depriving Hungary of its right

---

[327] It is worth noting that this case was the first case dealing with international watercourses to reach and be decided by the ICJ.

[328] The ICJ addressed the issue of "ecological necessity" at length, but inferred from the documents presented and discussed that "even if it had been established that there was in 1989, a state of ecological necessity linked to the performance of the 1977 Treaty, Hungary would not have been permitted to rely upon that state of necessity in order to justify its failure to comply with its treaty obligations, as it had helped, by act or omission to bring it about." *See supra* n. 326, paragraph 57.

[329] *See supra* n. 326, paragraph 85.

to an equitable and reasonable share of the natural resources of the Danube—with the continuing effects of the diversion of these waters on the ecology of the riparian area of the Szigetkoz—failed to respect the proportionality which is required by international law."[330] Thus, the decision of the ICJ clearly emphasizes the concept of equitable and reasonable utilization, and has made no reference to the obligation not to cause harm.

Situations may arise where the reasonable and beneficial uses of all the watercourse states cannot be fully realized because of existing utilizations by some riparians. The ILC addressed this issue and suggested that allocations to the other riparians would have to come from adjustments of the current uses. The utilizing riparians cannot refuse these adjustments on the basis that such adjustments would cause them harm.[331] Equality of rights of the riparians dictates the need for such adjustments.

It should be added that the determination that the principle of equitable utilization supersedes the obligation not to cause significant harm is not universally accepted. There are some experts in the field of international water law who still think that the principle of equitable and reasonable utilization and the obligation

---

[330] *See id*. The ICJ emphasized the principle of equitable and reasonable utilization further by linking it to the concept of equality of all the riparian states. It quoted from the 1929 judgment by the PCIJ where the PCIJ stated "[the] community of interest in a navigable river becomes the basis of a common legal right, the essential features of which are the perfect equality of all riparian States in the user of the whole course of the river and the exclusion of any preferential privilege of any one riparian State in relation to the others." *See River Öder Case, supra* n. 54. The ICJ went on to state that "Modern development of international law has strengthened this principle for non-navigational uses of international watercourses as well, as evidenced by the adoption of the Convention of 21 May 1997 on the Law of the Non-Navigational Uses of International Watercourses by the United Nations General Assembly." *See supra* n. 326, paragraph 85. The ICJ went further to clarify that certain actions by Hungary cannot be interpreted as forfeiture of Hungary's right to equitable and reasonable utilization. In this connection the ICJ stated: "It is true that Hungary, in concluding the 1977 Treaty had agreed to the damming of the Danube and the diversion of its waters into the bypass canal. But it was only in the context of a joint operation and a sharing of its benefits that Hungary had given its consent. The suspension and withdrawal of that consent constituted a violation of Hungary's legal obligations, demonstrating, as it did, the refusal by Hungary of joint operation; but that cannot mean that Hungary forfeited its basic right to an equitable and reasonable sharing of the resources of an international watercourse." *See id*., paragraph 78.

[331] On this issue, the ILC stated that ". . . where the quantity or quality of water is such that all the reasonable and beneficial uses of the all watercourse States cannot be fully realized, a 'conflict of uses' results. In such a case, international practice recognizes that some adjustments or accommodations are required in order to preserve each watercourse State's equality of right. These adjustments or accommodations are to be arrived at on the basis of equity, and can best be achieved on the basis of specific watercourse agreements." *See* Report of the International Law Commission on the Work of its Fortieth Session, 9 May—29 July 1988, (Supplement No. 10 A/43/10), United Nations 1988, at 84.

not to cause significant harm carry equal weight.[332] The Berlin Rules adopted by the ILA in 2004 also seem to equate the two principles.[333]

The Bank has not been shielded from this debate. It has taken the decision from its early years of involvement with international waters that it would not finance a project that could cause appreciable harm to another riparian without the consent of that riparian. This substantive rule was codified in the 1965 OM where management was required to report to the Executive Directors that "(i) the issues involved are covered by appropriate arrangements between the borrower and other riparians; or (ii) the other riparians have stated that they have no objection to the project; or (iii) the project is not harmful to the interests of other riparians and their absence of express consent is immaterial or their objections are not justified." Thus, from the start, the Bank clearly adopted the obligation not to cause harm as the substantive rule for its policy for projects on international waterways.

The policy elaborated the requirement that the project should not cause appreciable harm to other riparians in five separate areas:

First, it stated that the other riparians should be provided with the opportunity to determine if the project has the potential for causing appreciable harm to them through water deprivation, pollution or otherwise.[334]

Second, it indicated that the response of any of the riparians could be a confirmation that the project would not harm their interests.[335]

Third, the internal briefing memorandum from the staff to senior management on these responses required reporting all the facts including the staff assessment of whether the project would cause appreciable harm to the interests of other riparians, or be appreciably harmed by the other riparians' possible water use.[336]

Fourth, the project documents presented to the Executive Directors should deal with the international aspects of the project, and state that the Bank staff have considered these aspects and are satisfied that:

(a) the issues involved are covered by an appropriate agreement or arrangement between the beneficiary state and the other riparians; or

---

[332] *See* for example, Tanzi & Arcari, *supra* n. 185.

[333] *See supra* n. 172. For an overview of the Berlin Rules and how they dealt with the relationship between the two principles, *see* Salman, *The Helsinki Rules, supra* n. 172, at 625.

[334] *See* paragraph 3 of BP 7.50.

[335] *See id.,* paragraph 5.

[336] *See id.*

(b) the other riparians have given a positive response to the beneficiary state or Bank, in the form of consent, no-objection, support to the project, or confirmation that the project will not harm their interests; or

(c) in all other cases, in the assessment of Bank staff, the project will not cause appreciable harm to the other riparians, and will not be appreciably harmed by the other riparians' possible water use.

The cases referred to in (c) above would include the cases of objections, as well as the cases of exceptions to the notification requirement which will be discussed later.

Fifth, two of the exceptions to the notification requirement which will be discussed in the next Chapter are qualified by the requirement that the works that are exempted from notification should not "adversely change the quality or quantity of water flows to the other riparians"[337] or "cause appreciable harm to other states."[338]

The above requirements clearly indicate that the Bank policy for projects on international waterways has adopted and elaborated the obligation against causing "appreciable harm" to the other riparians, and required confirmation that the project would not cause such harm at the different stages of project processing. Those stages start with notification,[339] response to notification,[340] briefing to Bank management following completion of notification,[341] as well as in reporting to the Executive Directors.[342]

However, the policy does not define the term "appreciable harm" nor does it provide any guidance as to how it can be assessed. The obligation not to cause harm, or *sic utere tuo ut alienum non laedas,*[343] has been explained by Lucius Caflisch as follows:

> The no-harm rule probably originated from the consideration that, as in the case of neighboring owners of real property, neighboring States may not act as they please on their territories. They are not allowed to use or to tolerate the use of their territory for causing damage to their neighbors. This principle, which is linked to the concept of abuse of rights and which originated

---

[337] *See* paragraph 7(a) of OP 7.50.

[338] *See id.,* paragraph 7(c).

[339] *See* paragraph 3 of BP 7.50.

[340] *See id.,* paragraph 5.

[341] *See id.*

[342] *See* paragraph 8 of OP 7.50.

[343] *See supra* n. 67.

in the sphere of private law, appears to be a "general principle of law recognized by civilized nations" which, by now, has also entered the realm of customary international law.[344]

During the preparation of the UN Watercourses Convention, the ILC also struggled with the term, and took a while to reach a working definition. In 1985 the ILC stated that:

> The term "appreciable" embodies a factual standard. The harm must be capable of being established by objective evidence. There must be a real impairment of use, i.e. a detrimental impact of some consequence upon, for example public health, industry, property, agriculture, or the environment in the affected State. "Appreciable" harm is, therefore, that which is not insignificant or barely detectable but is not necessarily "serious."[345]

As discussed earlier, the decision in the *Trail Smelter* case gave some indications as to how the harm can occur. The tribunal in that case ruled that no state has the right to use or permit the use of its territory in such a manner as to cause injury by fumes in or to the territory of another or the properties or persons therein.[346]

Although the Bank policy has been crafted around the obligation not to cause harm, the Bank deliberated both principles (equitable and reasonable utilization and the obligation not to cause harm) during the preparation of the Memorandum on Riparian Rights of 1985.[347] The Memorandum devoted lengthy portions discussing both concepts. The difficulties of adopting the principle of equitable and reasonable utilization as the guiding principle of the Bank policy were amply articulated by Raj Krishna:

> The Bank policy could have been developed around the principle of equitable utilization also. The big hurdle here, however, is that for each project the Bank will need to determine whether the particular use to be financed by it falls within the equitable utilization of the beneficiary state, further necessitating an analysis of what is equitable for other riparians—a task the Bank cannot accomplish without the agreement of other coriparians as it is neither a court nor a tribunal. On the other hand, the no harm rule seemed

[344] *See* Caflisch, *supra* n. 56, at 12.

[345] *See* Report of the International Law Commission on the Work of its Fortieth Session, 9 May–29 July 1988, at 85.

[346] *See supra* n. 79.

[347] *See supra* n. 151.

more appropriate for the Bank and simple to apply. Moreover, avoidance of harm as the guiding principle is embedded in the Bank's practice over the years. Its adoption by the Bank in no way derogates from the significance of the principle of equitable utilization.[348]

It should, however, be emphasized that an absolute severance of the two principles is neither possible, nor practical, particularly in the case of the Bank policy for projects on international waterways. The principle of equitable and reasonable utilization deals with the quantitative sharing of the waters of the shared river, lake or aquifer. A project that allocates a disproportionate amount of water to one riparian at the expense of the other riparians will be considered by the Bank as causing appreciable harm to other riparians, and the Bank will not finance it.[349] Conversely, when a project uses a minimal amount of water from the shared river, lake, or aquifer, the Bank can determine, and indeed has been determining, that such use will not cause appreciable harm to the other riparians. In such cases, the Bank has proceeded with the financing of the project, even when there was an objection from one or more of the riparians. In both instances the Bank has made a determination on the amount of the water that may result in appreciable harm. In doing so, the Bank has invoked, wittingly or unwittingly, the principle of equitable and reasonable utilization, and made its determination accordingly. In the first instance the determination was that the project would cause appreciable harm because the amount of water to be abstracted by the project was disproportionate, and as such was not within the equitable and reasonable share of the riparian borrower. In the second case the determination was that the project would not cause appreciable harm to the other riparians because the quantity of the water it would use was minimal, and therefore within the ambit of the equitable and reasonable utilization of the borrower riparian.[350] As may be recalled, in the Igdir-Aksu Project in Turkey, the Bank made a determination that there would be a small reduction in energy output in Iran due to a minor decrease in the flow of the water of the Aras River in Iran. However, the use of this water by Turkey would constitute part of its reasonable and equitable share of the Aras River waters.[351]

This approach of linking the principle of equitable and reasonable utilization and the obligation not to cause harm applies to both downstream as well as

---

[348] *See* Krishna, *supra* n.105, at 36–7.

[349] The ICJ reached the same conclusion in the *Gabčikovo-Nagymaros* case when it ruled that the unilateral diversion by Slovakia of disproportionate amounts of the waters of the Danube River deprived Hungary of its right to an equitable and reasonable share of these waters. *See supra* n. 326.

[350] *See* Salman, *supra* n. 181, at 9.

[351] *See* Igdir-Aksu Project, *supra* n. 143.

upstream riparians, thus underscoring the need for all riparians be notified, as has been the policy and practice of the Bank. The approach also helps explain the concept of foreclosure of the future uses which will be discussed in more detail in Chapter 5, Part 5.3 of this Book. A project in a downstream riparian country that uses a disproportionate amount of water can cause appreciable harm to the other riparians because it will help the downstream riparian establish rights to the water acquired by the project, and the downstream riparian will thus be helped in claiming those water rights as established or acquired rights in any future negotiations. This in turn could preclude the upstream riparian or riparians from using those "foreclosed" waters because such use would, in the view of the downstream riparian, cause appreciable harm.

Clarifying the relationship between the two principles, Stephen McCaffrey stated that ". . .while the no-harm principle does qualify as an independent norm, it neither embodies an absolute standard nor supersedes the principle of equitable utilization where the two appear to conflict with each other. Instead, . . . it plays a complementary role, triggering discussions between the states concerned and perhaps, in effect, proscribing certain forms of serious harm."[352] Along the same lines, Ibrahim Shihata called the relationship between the two principles "the fictitious dichotomy." He cautioned that this fictitious dichotomy "need not stand in the way of such cooperative management for the optimal and sustainable uses of the international waterways. After all, equitable distribution must take account of existing uses and the need to maintain the livelihood of the population who came to be dependent on these uses."[353]

It is worth stressing that the policy of the Bank is based on the obligation not to cause appreciable harm to the other riparians. However, it should also be clarified that the policy of the Bank for all practical purposes has been a confluence point for the obligation against causing appreciable harm, and the principle of equitable and reasonable utilization. Through the practical application and implementation of the obligation against causing appreciable harm and the interface of this obligation with the principle of equitable and reasonable utilization, the policy has clearly demonstrated the relationship between the two concepts, and has clarified how they actually and practically interact.

---

[352] *See* McCaffrey, *supra* n. 185, at 408.

[353] *See* Ibrahim Shihata, "Foreword" to *International Watercourses—Enhancing Cooperation and Managing Conflict, supra* n. 56, at vii.

CHAPTER 5

The Notification Process

5.1 Basis of the Duty to Notify

The previous Chapter discussed the two elements on which the Bank policy for projects on international waterways is based. The first element is the recognition by the Bank that cooperation and goodwill of all riparians is essential for the efficient utilization and protection of the international waterway, and the readiness and willingness of the Bank to assist riparians in achieving such cooperative modalities. The second is the general rule that all riparians should be notified, subject to certain specified exceptions, of Bank-financed projects on international waterways.

The duty to notify other states of activities that may affect them stems from good faith, good neighborliness, and reciprocity.[354] It is an extension of the general obligation under international law to cooperate, and to exchange data and information on shared watercourses.[355] Such cooperation is no doubt the *sine qua non* for an efficient, equitable and sustainable utilization of international watercourses. In turn, notification is the only effective way for knowing about, and checking unilateral activities that may affect other riparian states.

A number of judicial decisions, conventions and treaties, as well as rules issued by the IIL and ILA, have dealt with the duty to notify. The arbitral tribunal

---

[354] *See* Maria Manuela Farrajota, *Notification and Consultation in the Law Applicable to International Watercourses,* in *Water Resources and International Law,* at 281 (Laurence Boisson de Chazournes & Salman M. A. Salman, eds., The Hague Academy of International Law 2005).

[355] Charles Bourne distinguished between three kinds of notice by a state to a co-basin state: "First, there is the notice that may be given by a state that is proposing to utilize the waters of a drainage basin; then there is the notice that may be given by a state that is aware of a situation in its territory, not voluntarily created by it, which may cause injury to co-basin states; and, finally, there is the notice that may be given by a state to warn co-basin states that their acts may affect its present or future rights or interests in the resources of the drainage basin." *See* Bourne, *International Water Law, supra* n. 71, at 144. This is a comprehensive presentation of the situations that should give rise to notification. The UN Watercourses Convention addresses the three cases as well. In addition to notification for planned measures which is dealt with under Articles 12–20, Article 28 deals with notification of emergency situations, while Article 18 deals with notification by a state warning another watercourse state that its planned measures may have significant adverse effects. However, for the purpose of the Bank policy, this Book will only be concerned with the first kind of notification.

in the *Lake Lanoux* case amply clarified the basis for notification in 1957 when it stated:

> A state wishing to do that which will affect an international watercourse cannot decide whether another State's interest will be affected; the other state is the sole judge of that and has the right to information on the proposals.[356]

The same rationale for the need for notification was used earlier in the Montevideo Declaration issued by the Seventh International Conference of American States in 1933. Paragraph 7 of the Declaration requires that "the works which a state plans to perform in international waters shall be previously announced to the other riparian or co-jurisdictional states. The announcement shall be accompanied by the necessary technical documentation in order that the other interested states may judge the scope of such works, and by the name of the technical expert or experts who are to deal, if necessary, with the international side of the matter."[357]

Among the early treaties that addressed the issue of notification is the Indus Waters Treaty concluded between India and Pakistan in 1960. Although the Treaty has divided the six rivers of the Indus System between the two countries,[358] it nevertheless requires the state that plans to construct engineering works that would cause interference with the waters of any of the rivers of the Indus Basin, and which in its opinion would affect the other party materially, to notify the other party of its plans.[359] The notifying state is also required to supply the other state with the available data relating to the proposed works to enable that state to inform itself of the nature, magnitude and effect of the works.[360]

---

[356] *See supra* n. 82, at 130.

[357] Paragraph 6 of the Montevideo Declaration deals with navigation and requires the riparian states not to carry any works that may injure free navigation, and to communicate to the other riparian states studies on any planned works. For the full text of the Declaration, *see* 28 Am. J. Int'l L., supp. 59 (1934).

[358] *See supra* n. 220.

[359] *See* Article VII(2) of the Indus Waters Treaty, *supra* n. 219.

[360] Article VII(2) of the Indus Waters Treaty goes on to state "If a work would cause interference with the waters of any of the rivers but would not, in the opinion of the Party planning it, affect the other Party materially, nevertheless the Party planning the work shall, on request, supply the other Party with such data regarding the nature, magnitude and effect, if any, of the work as may be available." Annexure D of the Treaty deals with plants and construction of river works for the generation of hydroelectric power by India on the Western Rivers allocated to Pakistan, and the provision of information on such plants and works by India to Pakistan, and Pakistan's response. Similar provisions are also included in Annexure E with regard to storage of waters by India on the Western Rivers.

Mention should also be made of the 1968 African Convention on the Conservation of Nature and Natural Resources. The Convention requires, "where surface and underground water resources are shared by two or more of the contracting States, the latter shall act in consultation, and if the need arise, set up inter-State Commissions to study and resolve problems arising from the joint use of these resources, and for the joint development and conservation thereof."[361] Thus, the Convention incorporates not only the principle of consultation but also, impliedly, the principles of notification and negotiations in good faith.

The 1973 Treaty between Uruguay and Argentina Concerning the Rio de la Plata and the Corresponding Maritime Boundary includes detailed provisions on notification.[362] Such provisions include description of the works, the time frame for response, type of response, the right of the notified party to inspect the works, and dispute resolution in case of differences. These provisions are reiterated in the 1975 Statute of the River Uruguay concluded between Argentina and Uruguay.[363]

A number of recent treaties, conventions and other instruments have dealt with the concept of notification and consultation. The Mekong Agreement, for example, defines notification as "the timely providing of information by a riparian to the Joint Committee on its proposed use of water according to the format, content and procedures set forth in the Rules for Water Utilization and Inter-Basin Diversions under Article 26."[364]

The 1992 United Nations Economic Commission for Europe (UNECE) Convention on the Protection and Use of Transboundary Watercourses and International Lakes (the Helsinki Convention) requires the parties to provide for the widest exchange of information, as early as possible, on issues covered

---

[361] *See* Article V(2) of the African Convention, FAO Legislative Study, *supra* n. 137, at 3.

[362] Article 17 of the Treaty states that "If one party plans to build new channels, substantially modify or alter existing ones or carry out other works, it shall notify the Administrative Commission which shall determine on a preliminary basis and within a maximum period of 30 days whether the plan might cause significant damage to the navigation interests of the other party or the regime of the river." For a copy of the Treaty, *see* 1295 U.N.T.S. 307 (1982). For discussion of the Treaty, *see* Lilian del Castillo-Laborde, *The Rio de la Plata and its Maritime Front Legal Regime* (Martinus Nijhoff Publishers 2008).

[363] *See* Statute of the River Uruguay, 1295 U.N.T.S. 340 (1982).

[364] *See* the Mekong Agreement, *supra* n. 242, Chapter II (Definition of Terms). Article 26 of the Agreement requires the Joint Committee to prepare and propose for approval of the Council, *inter alia*, Rules for Water Utilization and Inter-Basin Diversions. Those Rules have been prepared and approved by the Council in 2003, and they include "Procedures for Notification, Prior Consultation and Agreement." *See*: http://www.mrcmekong.org/download/programmes/PNPCA.pdf.

under its provisions.[365] Article 10 of the Helsinki Convention calls for consultations on the basis of reciprocity, good faith and good neighborliness. The UNECE 1991 Convention on Environmental Impact Assessment in a Transboundary Context (the Espoo Convention) requires notification by the party of origin[366] for any proposed activity listed in Appendix I to the Convention that is likely to cause another party significant adverse transboundary impact. The list of such impacts includes waste water disposal, large dams and reservoirs as well as groundwater abstraction.[367]

As discussed earlier, the IIL Salzburg Resolution requires each riparian state not to undertake works or utilization of the waters of a watercourse or hydrographic basin that would seriously affect the possibility of utilization of the same waters by other states except after previous notice to interested states. Similarly, the ILA Helsinki Rules issued in 1966 require a state, regardless of its location in a drainage basin, to furnish to any other basin state, the interests of which may be substantially affected, notice of any proposed construction or installation that would alter the regime of the basin in a way which might give rise to a dispute. Similarly, Principle 19 of the 1992 Rio Declaration on Environment and Development deals with prior and timely notification.[368]

The UN Watercourses Convention includes detailed provisions on planned measures. It requires the watercourses states to exchange information and consult each other, and if necessary, negotiate on the possible effects of planned measures

---

[365] *See* Article 6 of the Helsinki Convention, 31 I.L.M. 1312, (1992). *See* also: http://www.helcom.fi/Convention/en_GB/convention/.

[366] *See* Convention on Environmental Impact Assessment in a Transboundary Context, done February 25, 1991, 1989 U.N.T.S. 310 (1997): 30 I.L.M. 800 (1991) (hereinafter the "Espoo Convention"). *See* also: http://www.unece.org/env/eia/eia.htm. Article 3 of the Espoo Convention states: "1. For a proposed activity listed in Appendix I that is likely to cause a significant adverse transboundary impact, the Party of origin shall, for the purposes of ensuring adequate and effective consultations under Article 5, notify any Party which it considers may be an affected Party as early as possible and no later than when informing its own public about that proposed activity. 2. This notification shall contain, *inter alia*: (a) Information on the proposed activity, including any available information on its possible transboundary impact; (b) The nature of the possible decision; and (c) An indication of a reasonable time within which a response under paragraph 3 of this Article is required, taking into account the nature of the proposed activity; and may include the information set out in paragraph 5 of this Article."

[367] *See* paragraph 12 of Appendix I of the Espoo Convention (List of Activities).

[368] Principle 19 of the Rio Declaration reads: "State shall provide prior and timely notification and relevant information to potentially affected States on activities that may have a significant adverse transboundary environmental effect and shall consult with those States at an early stage and in good faith." For the Rio Declaration, *see* 31 I.L.M. 874 (1992); also included as Annex 6, in *Groundwater: Legal and Policy Perspectives*, World Bank Technical Paper No. 456, 139 (Salman M. A. Salman, ed., The World Bank 1999) at 257.

on the condition of the watercourse.[369] It also requires a watercourse state before it implements or permits implementation of planned measures which may have significant adverse effects upon the watercourse states to provide those states with timely notification. Such notification shall be accompanied by available technical data and information, including the results of any environmental impact assessment. As will be discussed later, the Convention has influenced a number of subsequently adopted treaties and conventions which include similar provisions on notification.

One common feature of all these instruments is that they set a threshold for notification. For the IIL, this threshold is that the works would seriously affect the possibility of utilization of the waters of the shared watercourse; for the ILA, the threshold is that the planned works would substantially affect the interests of other riparian states. The UN Watercourses Convention establishes "significant adverse effects" as the threshold for notification.[370] Attempting to clarify what the term "significant" means, the Working Group stated that:

> The term "significant" is not used in this article or elsewhere in the present Convention in the sense of "substantial." What is to be avoided are localized agreements, or agreements concerning a particular project, programme or use, which have a significant effect upon third Watercourse States. While such an effect must be capable of being established by objective evidence and not be trivial in nature, it need not rise to the level of being substantial.[371]

However, the Bank policy for projects on international waterways is again peculiar in this regard in that it does not set a threshold for notification. As discussed earlier, the policy specifies both, the types of waterways, as well as the projects to which it applies, and requires notification of the other riparians of the said project. Notification is required regardless of how insignificant, or *de minimis*, the effects of the project may be on the other riparians. Indeed, there may be cases where there are no effects whatsoever on some or on any of the riparians. Nonetheless, the policy sets forth the general rule of notification in those cases as

---

[369] *See* Article 11 of the UN Watercourses Convention.

[370] It should be noted that the Convention under Article 7, as discussed earlier, obliges the watercourse states not to cause "significant harm" while the threshold for notification under Article 12 of the Convention is "significant adverse effects."

[371] *See* Statements of Understanding issued by the Sixth Committee Convened as a Working Group of the Whole, (*see supra* n. 179). In earlier drafts of the Convention the threshold used was "appreciable adverse effects." *See* United Nations, Report of the International Law Commission on the Work of its Fortieth Session, 9 May—29 July, 1988 (Supplement No. 10 (A/43/10), New York 1988) at 79.

well. As will be discussed later, the policy allows for three exceptions to the notification requirement, but none of them relates to the threshold for notification.

The policy thus closed a gap in the interpretation of the pre-1985 directives that was used in some cases to allow proceeding with a project on an international waterway without notifying the other riparians if it was determined that the project would not be harmful to the interests of those riparians. That interpretation allowed the subjective determination of no impacts, and the subsequent lack of the need for notification.[372] However, henceforth any of the projects specified in the policy falling on any of the waterways under the policy would require notification of the other riparians, unless it is determined that the project is under one of the exceptions to the notification requirement discussed in Chapter 7 of this Book.

The basis for notification in the case of the Bank is more pressing than those of the other international instruments discussed above. In addition to the concepts of good faith, good neighborliness and reciprocity, the Bank, as an international financial cooperative institution, and its Articles of Agreement require it to act prudently in the interests of all its members. The Bank needs to keep the triangular relationship (between itself and the borrower; between itself and the other riparians, and among the riparians themselves) intact. Projects on international waterways may affect any and all of these relations, and hence there is need on the part of the Bank for extra caution. Moreover, and as the policy states, the cooperation and goodwill of all riparians is essential to the efficient utilization and protection of the waterway. This cooperation can only be realized through exchange of data and information, and notification.

For all those reasons, the Bank will not finance a project that will cause appreciable harm to other riparians. Following the *ratio decidendi* of the *Lake Lanoux* case, the Bank cannot, and indeed should not, make such determination for the

---

[372] David Goldberg, former Assistant (and later Deputy) General Counsel of the Bank, stated that in projects where there was "no issue, no conceivable harm," notification was not required by the Bank for projects on international waterways. *See* David Goldberg, *Legal Aspects of the World Bank Policy on Projects on International Waterways*, 7 International Journal of Water Resources Development 225, 228 (1991). What was meant, it seems, was that, if there is no conceivable harm, then there is no issue, and accordingly, no requirement to notify other riparians. However, this view is not supported by the provisions of the 1985 OMS, and the directives issued thereafter. Those policies set the general rule of notification, and included three exceptions. The exceptions are specific and do not lay down a general rule of "no conceivable harm, no notification" as suggested by David Goldberg. No memorandum was issued by senior management of the Bank (*supra* nn. 105 & 130) providing this interpretation. Moreover, neither of Raj Krishna's two articles cited in this Book makes a reference to such an exception or interpretation. The present author who has had a key advisory role in the application of the policy since 1997, is not aware of this exception or interpretation. Furthermore, the notion of "conceivable harm" was certainly something new and would have needed a clear definition to avoid subjective determination.

state. Hence, the policy established the requirement for notification, so as ". . . to enable the other riparians to determine as accurately as possible whether the proposed project has the potential for causing appreciable harm through water deprivation, pollution or otherwise."[373]

Another matter of concern to the Bank is that the project it plans to finance should not be appreciably harmed by the other riparians' possible water use. Notification has been thought of as one way of ascertaining other riparians' planned use to ensure that such use would not appreciably harm the project. Accordingly, the policy requires that the memorandum to Bank management on the responses to the notification, as well as the project's documents submitted to the Executive Directors, should confirm that the project will not be appreciably harmed by the other riparians' possible water use.[374] However, experience has shown that the competing interests and demands of the riparians over the shared waterway would make such an approach extremely difficult. No state would be willing to commit to another riparian or to a third party that it would use the waters of the shared waterway in such a way as not to appreciably harm a project in another riparian state. As a result, this element of the policy has gradually, through practice, been de-emphasized.

Thus, the above analysis shows that the basis for notification under the Bank policy is wider and more pressing because of the nature of the Bank as an international financial cooperative institution, and the requirements under its Articles of Agreement to act prudently in the interests of all its members.

5.2   Notification: By Whom?

The Bank policy for projects on international waterways requires the Bank to ensure that the international aspects of the project on an international waterway are dealt with at the earliest opportunity. When such a project is proposed, the Bank requires the beneficiary state (prospective borrower) to notify the other riparians of the project and provide them with the project details (Project Details). If the prospective borrower indicates to the Bank that it does not wish to give notification, the Bank can undertake such notification on behalf of the borrower, provided that the prospective borrower has no objection to the Bank doing so.[375]

---

[373] *See* paragraph 3 of BP 7.50.

[374] *See* paragraph 8 of OP 7.50, and paragraphs 5 and 6 of BP 7.50.

[375] The request from the borrower to the Bank to notify other riparians on its behalf has taken the form of a letter or an electronic mail from the representative of the borrower to this effect. In some instances this request was simply recorded in a project mission aide memoire.

In fact, the Bank has been requested in a large number of projects to give the notification itself, on behalf of the borrowers.[376] Some of the borrowers in these projects may not have good neighborly relations with some of the riparians. Other borrowers have taken note that some of the other riparians have carried out projects on the shared waterway, and did not notify them. Because of this failure or refusal to notify them, these borrowers feel that they are under no obligation to notify those riparians. This is reciprocity in a negative sense.

The willingness of the Bank from the early years to undertake notification on behalf of the borrower has helped fill a diplomatic vacuum, and brought in a third party as a go-between for the borrower and the other riparians. The ILC addressed the issue of notification by a third party as well, by including Article 30 in the UN Watercourses Convention on indirect procedures. This Article allows the riparians, in cases where there are serious obstacles to direct contacts between them, to carry out the notification, as well as exchange of data and information, through any indirect procedures acceptable to them.[377]

In some instances, the Bank has been asked to notify some riparians of the project, while the borrower was agreeable to notifying the other riparians of the same shared waterway about the same project. This request for splitting the notification process between the borrower and the Bank stems in most cases from the fact that the relationship of the borrower with some of the riparians may not be friendly. However, the Bank decided that because of the need to have a consistent approach, and furnish the same data and information to all the riparians, notification should be undertaken to all riparians either by the Bank or by the borrower, and that under no circumstances should it be undertaken partly by the Bank and partly by the borrower for the same project. In those types of circumstances, the borrower usually ends up asking the Bank to undertake notification to all the riparians on its behalf.[378]

---

[376] Some questions were raised as to who within the Bank, should be sending the notification letter: the country director of the notifying country, or the country director of the notified country (in cases where each country has a separate director)? The notification letter should go from the country director of the notifying country because that Country Department has all the information about the project. There may also be an element of conflict of responsibility if the letter is sent by the director of the notified country itself.

[377] Article 30 of the UN Watercourses Convention states that "In cases where there are serious obstacles to direct contacts between watercourse States, the States concerned shall fulfill their obligations of cooperation provided for in the present Convention, including exchange of data and information, notification, communication, consultations and negotiations, through any indirect procedure accepted by them."

[378] A question arose in one case where the riparians have concluded a treaty on navigation, and the proposed project included a component on navigation, as well as other components on water supply, all of which related to the shared river. Can the borrower undertake

In case the prospective borrower indicates to the Bank that it does not wish to give the notification, and objects to the Bank doing so on its behalf, the policy requires that the Bank to discontinue processing of the project.[379] The Bank will need to do so not only because of the need to comply with its policy for projects on international waterways, but also because of the requirement under its Articles of Agreement to act prudently in the interests both of the particular member in whose territories the project is located, and of the other members as a whole.[380] If the Bank were to proceed with project processing without notification of the other riparian states when such notification is required, it would have failed to comply with both, its Articles of Agreement, requiring the Bank to act prudently in the interests of all its members, and its policy for projects on international waterways.

It is worth adding that the prospective borrower that refuses to notify or allow the Bank to notify the other riparians of Bank-financed projects would have been, at one point or another, a beneficiary of the policy itself, because notification has been routinely undertaken for all riparians either by the borrowers or the Bank for a large number of Bank-financed projects on international waterways.[381] In the few instances when the Bank discontinues processing of the project because of the refusal of the prospective borrower to notify the other riparians, or to allow the Bank to do so, the policy requires the Bank to inform the Executive Directors concerned of these developments, and any further steps taken.[382]

However, there have been cases where the Bank-financed project was redesigned as a result of the refusal of the prospective borrower to undertake the notification or allow the Bank to undertake it on its behalf. The redesign of the project would drop the component or components that affect the international waterway, and have the Bank finance only the other components that have no international waterways aspects. This approach would, however, be adopted only

---

the notification for the navigation component under the said treaty, and ask the Bank to undertake notification on the other components? For the same reasons discussed above it was concluded that notification for all the components should be undertaken either by the borrower or the Bank.

[379] The ILA Helsinki Rules address this issue in Article XXIX(4) which states: "If a State has failed to give the notice referred to in paragraph 2 of this Article, the alteration by the State in the regime of the drainage basin shall not be given the weight normally accorded to temporal priority in use in the event of a determination of what is a reasonable and equitable share of the waters of the basin."

[380] *See supra* n. 84.

[381] Indeed, this argument has been used by the Bank effectively in a number of cases, and has helped relax opposition to notification by some prospective borrowers, who then asked the Bank to undertake the notification on their behalf.

[382] *See* paragraph 4 of OP 7.50.

if the project remains still viable without these components. It would allow the Bank to avoid non-compliance with its policy, and averts disputes with other riparians.[383] Another approach is limiting the components on the international waterway to those that can fit within the exception to the notification requirement dealing with the rehabilitation of existing schemes, as will be discussed in Chapter 7, Part 7.2.1 of this Book.

In addition to the prospective borrower and the Bank, notification may be undertaken by a third party. The Nile Basin Initiative (NBI) Secretariat has recently been undertaking notification of other riparians for Bank-financed projects. The Bank has agreed that the NBI Secretariat can undertake the notification to the other Nile riparian states for Bank-financed projects, following a request to this effect from the Nile Council of Ministers.[384] Under this arrangement, the NBI Secretariat acts on behalf of the borrower. It sends the notification to all the other riparians, and asks for any responses they may have. It coordinates its actions during the notification process with the borrower, as well as the Bank. The Bank agreed to this arrangement because it is felt that it would advance the process of the exchange of data and information and streamline the notification process among the Nile riparian states. It would also assist in strengthening the NBI Secretariat and enhancing the cooperative arrangements between the Nile states. Indeed, this approach would assist in achieving the Bank's objective of supporting the riparians' collaborative efforts to manage, share and protect the shared waterway, as well as support the objective of dispute avoidance.

Territories under the administration of the United Nations present special challenges with regard to notification. One such example is Kosovo while it was being administered by the United Nations Mission in Kosovo (UNMIK). The Bank decided that for such projects, the UNMIK would be approached and asked to undertake the notification on behalf of Kosovo. If UNMIK indicated to the Bank that it did not wish to do so, the Bank would undertake the notification provided UNMIK had no objection to the Bank doing so.[385] Undertaking notification through UNMIK was based on an earlier precedent where a similar United Nations administration was used. As will be discussed later, the United Nations

---

[383] Obviously, there is nothing that prevents the borrower from funding that component or components separately using its own resources. However, this could be said about the whole project if the Bank decides to discontinue processing it.

[384] For more information on the NBI and its different organs, *see supra* n. 238.

[385] One project that involved this arrangement was the Kosovo, Private Sector Participation in Water Supply and Sanitation in Gjakovë-Rahovec (P072913, 2002), which involved use of the water of the Drin/Drim River that Kosovo shares with Albania. The UNMIK agreed to send the notification about the project to Albania, to which Albania responded positively.

Council for South West Africa (Namibia) was notified of a project in Lesotho that involved use of the waters of the Orange River that Namibia shares with Lesotho and South Africa.[386]

As will be discussed in the next Part of this Chapter, some of the river basin organizations that are being notified of a Bank-financed project under the respective legal instrument establishing them may act as recipients of this notification on behalf of the riparian members of the organization. The river basin organization could simply pass on the notification to the riparian states, and compile their responses and send them to the notifying state or the Bank (if the Bank has undertaken the notification). Under this arrangement, the river basin organization simply passes on the notification, whereas under the NBI arrangements, the notification originates from the NBI on behalf of the borrower.[387]

As such, in addition to notification being undertaken by the borrower and the Bank, other innovative approaches are evolving. These approaches have assisted in accommodating new situations and expanding the Bank practice and precedents in this field.

5.3   Notification: To Whom?

One of the fundamental misunderstandings about international water law is that harm can only be caused by upstream riparians to the downstream riparians. Based on this faulty belief, it is widely thought by a large segment of both, lawyers and non-lawyers, that notification is an exclusive right of downstream riparians, because only upstream riparians can harm downstream riparians. In other words, harm can only "travel" downstream with the flow of the waters, and accordingly, downstream riparians do not need to notify upstream riparians of any activity that they are undertaking or plan to undertake because such activity cannot possibly harm the upstream riparians.

It is clear and obvious that the downstream riparians can be harmed by the physical impacts of water quantity and quality changes caused by the use of the shared waterway by the upstream riparians. The quantity of water flow can be decreased by upstream riparians through construction of dams and canals, and the storage and diversion of the waters of the shared river. The quality of the water of the

---

[386] *See* discussion of the Lesotho Highlands Water Project, *infra* n. 429.

[387] In the Eastern Nile Watershed Management Project (P111330, 2009), the NBI Secretariat notified, on behalf of Egypt and Sudan, the other eight Nile Basin riparian countries (Burundi, the Democratic Republic of Congo, Eritrea, Ethiopia, Kenya, Rwanda, Tanzania and Uganda). Similarly, NBI undertook the notification for the Egypt, West Delta Water Conservation and Irrigation Rehabilitation Project (P087970, 2007); as well as the Ethiopia, Irrigation and Drainage Project (P092353, 2007). In the latter two projects, all the other nine riparians were notified of the project by the NBI Secretariat.

shared river can also be affected by the upstream riparian through pollution caused by industrial waste, sewerage or agricultural run-off.[388]

It is much less obvious, and generally not recognized, that the upstream riparians can be affected, or even harmed, by the potential foreclosure of their future use of water, caused by the prior use and the subsequent claim of rights to such water by the downstream riparian. Projects on shared rivers in the downstream riparian states would help those riparians in acquiring rights to the water abstracted as a result of those projects, and the right to use of such waters in future by the other riparians would have been foreclosed by the downstream riparians. Such downstream riparians would usually invoke the principle of acquired rights, and the obligation not to cause harm, if and when the upstream riparians claim some rights to the foreclosed waters under the principle of equitable and reasonable utilization. As discussed earlier, this is one area where the two principles may appear to conflict.

Under this concept, a poor upstream country could be precluded from developing the water resources of a shared river in the near future if a richer downstream riparian, without consultation or notification, develops it at present. The downstream state would establish rights to those waters which they would claim as acquired rights, thus foreclosing the future uses of those waters by the upstream riparians.[389]

Although the concept of foreclosure of future uses is not widely comprehended as discussed earlier,[390] there have been cases where some upstream riparians have shown understanding of the concept and its implications for them.[391]

---

[388] It should be added that the downstream riparian can also harm the upstream riparian by constructing dams at the borders with the upstream riparian without the consent of that state. Such a dam could submerge lands in the upstream riparian, causing harm to that state. This was the case of the Aswan High Dam Project where the Bank insisted that Egypt reach an agreement with the Sudan, whose lands would be submerged, before the Bank would consider financing the Dam. As discussed earlier, Sudan consented to the construction of the Aswan High Dam and the submergence of some of its lands through the 1959 Nile Agreement with Egypt. *See supra* n. 132.

[389] *See* Stephen McCaffrey, *Water Disputes Defined: Characteristics and Trends for Resolving Them,* in *Resolution of International Water Disputes,* 59 (The International Bureau of the Permanent Court of Arbitration, ed., Kluwer Law International 2003). *See* also Salman, *supra* n. 181.

[390] *See supra*, Chapter 4, Part 4.5 of this Book.

[391] One such example is Ethiopia's Note Verbale of March 20, 1997, addressed to Egypt, and copied to the Mr. Kofi Annan, Secretary General of the United Nations; Mr. Salim Ahmed Salim, Secretary General of the Organization of the African Unity (OAU); and Mr. James Wolfensohn, President of the World Bank. The Note, as quoted by John Waterbury, stated: "Ethiopia wishes to be on record as having made it unambiguously clear that it will not allow its share to the Nile waters to be affected by a *fait accompli* such as the Toshka project, regarding which it was neither consulted nor alerted." *See* Waterbury, *The Nile Basin, supra* n. 234, at 84–5 (Yale University Press, 2000). The Toshka project involved

Some recent agreements on shared rivers have even included a reference to this concept.[392]

As mentioned before, one of the early approaches to projects on international waterways deliberated by the Bank in the 1950s was recognition of established uses of the shared river and protection of such uses against upstream withdrawals and downstream establishment of prior rights. This approach indicated that some understanding existed even in the 1950s of the fact that downstream riparians can also affect the upstream riparians. Although this notion was not clarified in any Bank directive, the way the requirement of notification was codified did not distinguish between downstream and upstream riparians. Starting in 1985 when notification was explicitly codified for the first time in OMS 2.32, up until today, the policy states that the prospective borrower ". . . should formally notify the other riparians of the proposed project."[393] The obligation to notify the other riparians has since 1985 been consistently applied to "all other riparians" and not only to the downstream riparians.

Indeed, when OD 7.50 was issued in 1989 referring to adverse effects and appreciable harm to downstream riparians, it was immediately withdrawn and reissued, replacing the terms" downstream riparians" and "upstream riparians" throughout the OD with the term "other riparians"[394] That correction emphasized and underscored the requirement under the policy that both downstream as well as upstream riparians be notified of the project. It is not clear whether the Bank followed this approach because it recognized the possibility of harm to both riparians, or simply to keep the 1989 OD consistent with the previous directives and

---

the irrigation of about 200,000 hectares in the south western desert of Egypt through a pumping station and canal that would convey water from the Aswan High Dam. *See id.* Another example is the 1961 letter from Prime Minister Nehru to President Ayub Khan which stated: "One more matter to which I must also refer, is the distinction you still seem to make between the rights of upper and lower riparians in paragraph 7 of your letter, which implies that the lower riparians can proceed unilaterally with projects, while the upper riparian should not be free to do so. If this was to be so, it would enable the lower riparian to create, unilaterally, historic rights in its favor and go on inflating them at its discretion thereby completely blocking all development and uses of the upper riparian. We cannot, obviously, accept this point of view." *See* Ben Crow *et al., Sharing the Ganges: The Politics and Technology of River Development*, 89 (Sage Publications, 1995).

[392] For example, the Senegal River Water Charter, which was concluded by Mali, Mauritania and Senegal in May 2002, and which Guinea signed in 2006, enumerates in Article 4, a number of principles for allocation of the waters of the Senegal River. Such principles include "the obligation of each riparian state to inform other riparian states before engaging in any activity or project likely *to have an impact on water availability, and/or the possibility to implement future projects.*" (Italics added). *See* also *supra* n. 239.

[393] *See* paragraph 5 of OMS 2.32 of 1985. This same language was used in all the succeeding directives.

[394] The relevant paragraphs of the OD were 8(a) on the exception to notification, and 11(c) on the staff assessment. The term "upstream riparians" in those two paragraphs was also replaced by "all riparians."

other rules and resolutions of other international organizations which used the term "other riparians."[395]

The different IIL and ILA rules discussed earlier also do not restrict notification to downstream riparians only. In fact the IIL Salzburg Resolution states that no state can undertake works or utilization of the waters of a watercourse or hydrographic basin that seriously affect the possibility of utilization of the same waters by other states except after previous notice to the interested states.[396] The ILA showed an understanding of this issue as early as 1964 during the Tokyo Conference when it concluded that "Any use of water by a riparian State, whether upper or lower, that denies an equitable sharing of uses by a co-riparian State conflicts with the community of interests of all riparian States in obtaining maximum benefits from the common source."[397] Following the same approach, the Helsinki Rules indicate that a state, regardless of its location in a drainage basin, should furnish to any other basin state, the interests of which may be substantially affected, notice of any proposed construction or installation which would alter the regime of the basin in a way which might give rise to a dispute.[398]

Furthermore, the UN Watercourses Convention requires that before a watercourse state implements or permits the implementation of planned measures which may have a significant adverse effect upon other watercourse states, it shall provide those states with timely notification thereof.[399] As such, the IIL and the UN Watercourses Convention have used the term "other states." The Helsinki Rules are even clearer as they require notification from any riparian state "regardless of its location in a drainage basin." Indeed, none of these instruments has made a distinction between downstream and upstream riparians when it comes to the requirement of notification.

Consistent with those instruments, the Bank has always emphasized the equal treatment of all the riparians, and its successive directives referred to effects of the projects it finances on "other riparians;" not just on downstream riparians.[400]

---

[395] There is also the view that the terms "downstream and upstream riparians" ignore the fact that a number of rivers are actually boundary rivers where there are no upstream and downstream riparians. *See supra* n. 51.

[396] *See* Articles 3 and 5 of the Salzburg Resolution, *supra* n. 69.

[397] *See* ILA, Tokyo Report 1964, at 165.

[398] *See* Article XXIX(2) of the Helsinki Rules, *supra* n. 166.

[399] *See* Article 12 of the UN Watercourses Convention, *supra* n. 180.

[400] One erroneous presentation of the Bank policy appears in Surya Subedi, *Resolution of International Water Disputes: Challenges for the 21st Century*, in *Resolution of International Water Disputes* (International Bureau of the Permanent Court of Arbitration), *supra* n. 389, at 33. The author, after quoting paragraph 2 of BP 7.50 on the requirement to "formally notify the other riparians. . ." stated that: "Thus, the guidelines of the World Bank

Based on this approach, the Bank has required notification of all the riparians of the shared waterway of any Bank-financed project that does not fall under one of the exceptions to the notification requirement discussed in Chapter 7 of this Book.

By way of examples, the Bank has required notification of all the riparians of the Mekong River (China, Myanmar, Lao PDR, Thailand, Vietnam and Cambodia) of any project on the Mekong or any of its tributaries in any of those countries.[401] Similarly, all the riparians of the Ganges or Brahmaputra are notified of any project on their respective river or its tributaries.[402] Along the same line, any project on the Nile would require notification of all the other Nile River riparian states (Burundi, the Democratic Republic of Congo, Egypt, Eritrea, Ethiopia, Kenya, Rwanda, Sudan, Tanzania and Uganda). No distinction has been made between downstream and upstream riparians, or between projects in the Eastern Nile countries (Ethiopia, Eritrea, Egypt and Sudan), or the Southern Nile countries

---

in effect require an upper riparian state to notify the lower riparian state of its intention to implement water projects upstream. When notified by the upper riparian state, the lower riparian state can object to the project on the grounds that the project will have an adverse impact on it. Once there is an objection from the lower riparian state, the World Bank is unlikely to finance the project. Hence, this policy of the Bank in effect grants a veto power to the lower riparian state to the detriment of the upper riparian state." *See id.*, at 37. As the discussion throughout this Book indicates, notification is required for all the riparians (downstream as well as upstream), and no riparian has veto power over a Bank-financed project.

[401] Under the Lao PDR, Khammouane Development Project (P087716, 2008), some proposed activities would support the rehabilitation of the existing irrigation schemes as well as development of some new small irrigation schemes of less than 50 hectares each. These activities are situated within the catchment area of the Xe Bang Fai River, which is a tributary of the Mekong River. The other five riparians of the Mekong River, both downstream and upstream, were all notified of the project, and there was no unfavorable response from any of them.

[402] The Nepal, Power Development Project (P043311, 2003) involved the development of two medium size (no more than 30 MW each), and one small hydropower sub-projects (under 10 MW) on the Gandaki and Kosi rivers. These rivers are tributaries of the Ganges River shared by Bangladesh, China, India and Nepal. It was determined that the "Revised Agreement on the Kosi Project" signed on December 19, 1966 between India and Nepal, as well as the "Agreement on Gandak Irrigation and Power Project" concluded on December 4, 1959 between India and Nepal, are project specific, and do not relate to the proposed Power Development Project. Accordingly, the World Bank, on behalf of Nepal, notified China (upstream riparian), and India and Bangladesh (downstream riparians) of the project, and no unfavorable response was received. For copies of each of the Kosi and Gandak Agreements, *see* Salman & Uprety, *supra* n. 61, at 349 & 361, respectively. Similarly, the India, Vishnugad Pipalkoti Hydro Electric Project (P096124, 2009) is planned as a 444 MW run-of-river power plant for which a 65-meter-high dam will be built. The project is situated on the Alaknanda River in the state of Uttaranchal, India. The Alaknanda River is a tributary of the Ganges River. The Bank, on behalf of India, notified all the other riparians of the Ganges River (China, Nepal and Bangladesh) of the project.

(Burundi, Democratic Republic of Congo, Kenya, Rwanda, Tanzania and Uganda).[403] Projects on the Niger River Basin in Nigeria, which is the lowest downstream riparian of the basin, have been notified to all the other riparians of the Niger Basin, or to the NBA.[404]

As explained earlier, notification can be undertaken either by the borrower or the Bank. If notification is undertaken by the Bank, it would be sent to the Ministry of Finance of each of the riparian states because the Ministry of Finance in each country is the representative of the borrower, and is the focal point for the Bank work and correspondence in each country. On the other hand, riparian states usually deal with each other through their Ministries of Foreign Affairs, because dealings with other states are considered matters of international and diplomatic relations.

The policy for projects on international waterways requires the Bank to ascertain whether the riparians have entered into agreements or arrangements, or have established any institutional framework for the waterway concerned. In such cases the Bank ascertains the scope of the institution's activities and functions and the status of its involvement in the proposed project, bearing in mind the possible need for notifying the institution.[405] Indeed, a number of treaties have established river or lake basin management organizations as legal entities and required the riparians to notify those organizations of any proposed works on such shared river or lake. One of the earliest treaties to deal with this matter is the 1964 Convention and Statutes Relating to the Development of the Chad Basin.[406] The Convention established the Lake Chad Basin Commission (LCBC, or the Commission), and entrusted the LCBC with wide authority. The Statutes state that:

> The Member States undertake to refrain from adopting, without referring to the Commission beforehand, any measures likely to exert a marked influence either upon the extent of water losses, or upon the form of the annual

---

[403] The Ethiopia, Tana and Beles Integrated Water Resources Development Project (P096323, 2008) included small-scale investments in watershed management and community-based flood management. The watershed activities covered an area of around 80,000 hectares in selected microcatchments of Lake Tana catchments. All the other nine riparians of the Nile Basin were notified of the project, and no unfavorable response was received. Similarly, the Egypt, West Delta Water Conservation and Irrigation Rehabilitation Project proposed to develop and irrigate an area between 24,000 and 36,000 hectares located approximately 60 kilometers northwest of Cairo. Because water will be conveyed to the project from the Nile, the other nine riparians of the Nile were all notified of the project, and no unfavorable response was received.

[404] *See* Nigeria, Second National Fadama Development Project (P063622, 2003) and Nigeria, Third National Fadama Development Project (Fadama III) (P096572, 2009).

[405] *See* paragraph 5 of OP 7.50.

[406] For the Convention and Statutes, *see supra* n. 137.

hydrograph and limnograph and certain other characteristics of the Lake, upon the conditions of their exploitation by other bordering States, upon the sanitary condition of the water resources or upon the biological character- istics of the fauna and flora of the Basin. In particular, the Member States agree not to undertake in that part of the Basin falling within their jurisdic- tion any work likely to have a marked influence upon the system of the water courses and level of the Basin without adequate notice and prior con- sultation with the Commission.[407]

The Bank has taken note of this requirement under the Convention and Statutes, and has been regularly notifying the LCBC of any projects on or likely to affect the Lake Chad Basin. One example was the Chad-Cameroon Petroleum Development and Pipeline Project of which the LCBC was notified of the project, in addition to the notification sent to the riparian states them- selves.[408] Similarly, the LCBC was notified of a number of Bank-financed projects in Nigeria, including the Fadama projects, discussed in more details later.[409] In turn, the LCBC has been responding promptly and appropriately to those notifications.

As indicated earlier, both the 1973 Treaty and the 1975 Treaty between Uruguay and Argentina on the Uruguay River requires that notification regarding the construction or modification of channels or other works affecting the River be given to the Administrative Commission established under the Treaty.[410] Similarly, the 1980 Convention Creating the Niger Basin Authority (NBA) obliges the nine riparians of the Niger River to keep the Executive Secretary of the NBA informed of all projects and works that they intend to carry out in the basin.[411] The Agree- ment on the Establishment of the Zambezi Watercourse Commission obliges each of the parties to notify the Commission of any activity on the Zambezi that may adversely affect the watercourse or any member state, and to provide the Commission with all available data and information with regard thereto.[412]

---

[407] *See* Article 5 of the Statutes.

[408] For discussion of this project and the notification process, *see supra* nn. 281 & 283.

[409] *See infra* nn. 647 & 648.

[410] *See supra* nn. 362 & 363.

[411] Article 4(4) of the Niger Convention states that: "The member States pledge to keep the Executive Secretariat informed of all the projects and works they might intend to carry out in the Basin. Moreover, they pledge not to undertake any work on the portion of the River, its tributaries and sub-tributaries under their territorial jurisdiction which pollute the waters or modify the biological features of the fauna and the flora." *See* Convention Creating the NBA, *supra* n. 247.

[412] *See* Article 16 of the Agreement on the Establishment of the Zambezi Watercourse Commission, *supra* n. 267.

The International Commission for the Protection of the Danube River (ICPDR, or the Danube Commission), established under the 1994 Convention on Co-operation for the Protection and Sustainable Use of the River Danube (the Danube Convention), requires that the exchange of information and consultation on planned activities be carried out within the framework of the ICPDR.[413] The Framework Convention for the Protection of the Marine Environment of the Caspian Sea[414] sets forth similar arrangements. One of the functions of the Secretariat of the Caspian Sea Convention, which is based in Tehran, is to prepare and transmit to the Contracting Parties notifications, reports and other information received. The ICPDR and the Caspian Sea Secretariat have both been notified of a number of Bank-financed projects that may affect the Danube River[415] and the Caspian Sea, respectively.[416]

The 1995 Convention for the Protection of the Marine Environment and the Coastal Region of the Mediterranean (also known as the Barcelona or Mediterranean Convention) has the interesting feature of designating an organization not established by the Convention as responsible for some secretarial functions. Article 17 of the Convention designates the UNEP to carry out those functions which include handling of notification.[417]

In these and similar cases, notification is sent (whether by the borrower or the Bank) to the organization specified in the respective agreement. That

---

[413] Article 11 of the Danube Convention states that: "Having had a prior exchange of information, the Contracting Parties involved shall, at the request of one or several Contracting Parties concerned, enter into consultations on planned activities as referred to in Article 3, paragraph 2, which are likely to cause transboundary impacts, as far as this exchange of information and these consultations are not yet covered by bilateral or other international cooperation. The consultations are carried out as a rule in the framework of the International Commission, with the aim to achieve a solution." The Danube Convention was signed on June 29, 1994, in Sofia, Bulgaria, by 11 of the Danube Riparian States (Austria, Bulgaria, Croatia, the Czech Republic, Germany, Hungary, Moldova, Romania, Slovakia, Slovenia and Ukraine, as well as the European Community), and duly came into force in October1998, when it was ratified by the ninth signatory. For more on the Danube Convention see http://www.icpdr.org/.

[414] *See supra* n. 253.

[415] The International Commission for the Protection of the Danube River (ICPDR) was notified of a number of Bank-financed projects, including Romania, Municipal Services Project (P088252, 2006), as well as Bulgaria, Municipal Infrastructure Development Project (P099895, 2009).

[416] The Caspian Sea Secretariat has been notified of a number of Bank-financed projects, including Azerbaijan, Rural Investment Project (P076234, 2004).

[417] For the Convention *see*: http://eelink.net/~asilwildlife/medmarine.html. Article 17 of the Convention states that: "The Contracting Parties designate the United Nations Environment Programme as responsible for carrying out the following secretarial functions . . . (ii) To submit to the Contracting Parties notifications, reports and other information received in accordance with articles 3, 9 and 26."

agreement and the organization's internal rules would determine whether the organization can make the decision on the notification itself, or it needs to refer the matter to the member countries. As we have seen, the Administrative Commission which was established by Uruguay and Argentina under the 1973 Treaty, and the 1975 Statute, has authority to determine on a preliminary basis, and within a maximum period of 30 days, whether the plan might cause significant damage to the navigational interests of the other party, or the regime of the river.

Borrowers usually follow these arrangements as well, and notify such river basin organizations when the borrowers are members of such organizations. In cases when the borrower is not a party to the treaty establishing the river basin organization, then the borrower would not be expected to send the notification to this organization, as discussed below. In some cases, such a river basin organization may actually be established by a national law, and not as a result of a multilateral or bilateral treaty.[418]

In some projects the Bank went the extra step and notified both the river basin organization as well as the riparian states, although notifying one or the other would have met the policy requirement. The Odra River Basin Flood Protection Project in Poland involved the construction of a polder and embankment across the Odra River.[419] The River originates in the Czech Republic, flows through Poland, and then forms the border between Poland and Germany before emptying into the Baltic Sea. The three parties established the International Commission for the Protection of the Odra/Oder against Pollution (ICPOAP) in

---

[418] The South African, National Water Act (Act No. 36 of 1998) includes a full chapter (Chapter 10) on International Water Management. Under said Chapter the Minister of Water Affairs and Forestry may establish bodies to implement international agreements regarding the management and development of water resources shared with neighboring countries, and on regional co-operation over water resources. The governance, powers and duties of these bodies are determined by the Minister in accordance with the relevant international agreement, but they may also be given additional functions, and they may perform their functions outside the Republic of South Africa. Certain existing international bodies such as the Komati Basin Water Authority established by an agreement dated March 13, 1992 with the Kingdom of Swaziland, and the Vioolsdrift Noordoewer Joint Irrigation Authority established by an agreement dated September 14, 1992 with the Government of Namibia, are deemed to be bodies established under the Act. This is the only national water law that makes such extensive and detailed provisions on the issue of international water management. For further discussion of this issue, *see* Robyn Stein, *South Africa's New Democratic Water Legislation: National Government's Role as a Public Trustee in Dam Building and Management Activities*, 18 Journal of Energy and Natural Resources Law 284 (2000). *See* also Salman M. A. Salman & Daniel Bradlow, *Regulatory Frameworks for Water Resources Management—A Comparative Study*, at 147 (Law, Justice, and Development Series, The World Bank 2006).

[419] *See* Poland, Odra River Basin Flood Protection Project (P086768, 2007). The river is also known as the "Oder River." *See supra* n. 54.

1996.[420] The Bank, at the request of Poland, notified the ICPOAP, as well as Germany and the Czech Republic. Although both Germany and the Czech Republic responded positively, the latter advised the Bank to deal also with the ICPOAP directly.

As stated earlier, some of the river basin organizations may not include all the riparian states of the shared waterway. Should notification in these kinds of situations be sent to the organization and the states that are not parties to the organization, or should it be sent to each of the riparians separately? A case in point is the Joint Committee which is part of the Mekong River Commission (MRC) established under the 1995 Agreement on the Cooperation for the Sustainable Development of the Mekong River Basin (the Mekong Agreement).[421] The Mekong Agreement includes detailed provisions on notification.[422] However, as explained earlier, only Cambodia, Lao PDR, Thailand and Vietnam are parties to the Agreement. China and Myanmar are not parties to the Agreement, although they have been termed "dialogue partners."[423] The question that arose in this and other similar situations is to whom should the notification be sent? The answer would depend on the situation of each project. If the borrower is one of the parties to the agreement or treaty, then under the agreement, notification would be sent to the river basin organization and each of the other non-members of the organization. The Bank would follow a similar approach if it is undertaking the notification on behalf of the borrower under the same situation.[424] If the borrower is not a member of the organization, then notification would be sent only to each of the other riparians, because the borrower is not party to the agreement, and thus would not use the mechanisms specified therein.[425]

---

[420] The Agreement was concluded by Poland, the Czech Republic, Germany and the European Union on April 1, 1996, in Wroclaw. The Agreement entered into force on April 28, 1999. For more details on the Odra/Oder River and the Agreement see A. Dubicki & A. Nalberczynski, *Development of the International Cooperation in the Oder River Basin,* in *Management of Transboundary Waters in Europe,* 177 (Malgorzata Landsberg-Uczciwek, Martin Adriaanse & Rainer Enderlein, eds., 1998).

[421] The Mekong River Commission (MRC) has been established under Article 11 of the Agreement, and consists of the Council, the Joint Committee and the Secretariat. *See supra* n. 242.

[422] For discussion of those provisions, *see supra* n. 364.

[423] *See supra* n. 245.

[424] In Nam Theun 2 Power Project (NT2) (P076445, 2005), the government of Lao PDR notified each of the other five riparians of the Mekong River, as well as the MRC, of the project. Since the members of the MRC were themselves notified, the notification to the MRC took the form of information, rather than actual notification.

[425] If the validity of the treaty itself is challenged by some of the riparians, then the Bank would process the notification through its policy, and not through the requirements of that treaty.

One other question that has arisen is whether recognition of the river basin organization established under the treaty would extend to the threshold for notification set forth in that treaty in lieu of the Bank's own threshold. Specifically, some treaties have established a certain threshold therein for consultation, the exchange of date and information, and notification. The 1994 Danube Convention, as well as the 2000 SADC Protocol, both establish the threshold of "significant adverse effects" for notification.[426] In such cases, should the Bank require notification for any project on international waterways as per its policy, or should it follow the provisions of these and similar agreements and limit notification only to projects that may cause significant adverse effects, as per the agreement? The Bank has consistently decided that in all such cases, it would follow the more strict provisions of its own policy, and not the provisions of such agreements. The main reasons given are the need to adhere to consistent standards, and the concern that not all such agreements are inclusive of the riparian states of the waterway in question. The latter situation could result in two standards for notification: one under the agreement, and the other under the policy. Hence, the Bank uses its own threshold and standards for notification.

Despite the details laid down in the policy on notification, the Bank occasionally faces some unusual cases of what organizations and states should be notified. The Emergency Irrigation Rehabilitation Project in Afghanistan[427] aimed at rehabilitation of some of the irrigation schemes that draw water from a number of rivers, including the Amu Darya River. The River is shared by Tajikistan, Afghanistan, Turkmenistan and Uzbekistan, and flows into the Aral Sea to which Uzbekistan and Kazakhstan are the riparian states. There was no disagreement that the Amu Darya riparian states should be notified. The idea of notifying Kazakhstan was resisted on the ground that Kazakhstan is not a riparian of the Amu Darya, the river from which the water would be used. It was further argued that the borrower, Afghanistan, is not a party to any of the treaties on the Aral Sea or the Amu Darya, and thus is under no obligation to notify Kazakhstan.[428] However, it was clarified, and finally agreed, that the Amu Darya is a tributary of the Aral Sea, and that any decrease in its water flow to the Aral Sea could adversely affect Kazakhstan, as well as the Aral Sea.

The Bank would have proceeded accordingly, and would have notified the Amu Darya riparian states as well as Kazakhstan. However, the project was

---

[426] *See* Article 11 of the Danube Convention, and Article 4(1)(b) of the SADC Protocol.

[427] Afghanistan, Emergency Irrigation Rehabilitation Project (P078936, 2003).

[428] *See* the 1992 Agreement between the Republic of Kazakhstan, the Kyrgyz Republic, the Republic of Tajikistan, Turkmenistan and the Republic of Uzbekistan on co-operation in interstate water resources use, protection and common management, available at: http://www.icwc-aral.uz/legal_framework.htm. Afghanistan is not a party to this Agreement despite the fact it is a riparian of the Amu Darya River.

finally processed as dealing with rehabilitation of existing schemes, thus falling under the exception to the notification requirement under paragraph 7(a) of the Bank policy, as will be discussed later. The confusion, and perhaps intricacy of the situation, might have been caused by the treatment of the Amu Darya as a tributary of the Aral Sea. It might also have arisen due to lack of reciprocity between Afghanistan and Kazakhstan on these two systems of international waterways. Kazakhstan would be under no obligation under international law generally, or the Bank policy, to notify Afghanistan of any works it plans on the Aral sea. On the other hand, Afghanistan is under such obligation to notify Kazakhstan for projects on the Amu Darya because it is a tributary of the Aral Sea.

The Lesotho Highlands Water Project presented another challenge regarding whom to notify. This is a multipurpose project that involved the construction of dams and tunnels and the development and transfer of water resources from the highlands region of Lesotho, to the Republic of South Africa, for municipal and industrial use.[429] The water to be exported to South Africa is from the Sengue River which originates in Lesotho.[430] The River becomes the Orange River upon entry into South Africa, and thereafter forms the boundary between South Africa and Namibia before emptying into the Atlantic Ocean. The riparian states of the River are therefore, Lesotho, South Africa and Namibia.

The issue of notification came up in the mid-1980s while Namibia was still under the administration of the United Nations Council for South West Africa (Namibia) (UN Council). The UN Council was established under a UN General Assembly Resolution in 1967.[431] The question therefore was how to notify Namibia.[432] It was decided to send the notification to the UN Council which was

---

[429] *See* Lesotho Highlands Water Project—Phase 1A (P001396, 1991). This Phase involved construction of the Katse Dam on the Sengue River, a transfer tunnel, delivery tunnel, and an underground power station. Phase 1B of the project (P001409, 1998) included the construction of the Mohale Dam on the Sengue River, as well as a tunnel and other diversions. The Sengue River is also known as the Senqu River.

[430] The basis of the Lesotho Highlands Water Project—Phase 1A (P001396, 1991) is the treaty concluded between the Kingdom of Lesotho and the Republic of South Africa on October 24, 1986, which laid down detailed provisions on the project, including the payment of royalties to Lesotho by South Africa. Project preparation commenced thereafter and the project was approved by the Bank in 1991. For a copy of the Treaty *see* FAO, Treaties Concerning the Non-Navigational Uses of International Watercourses, *supra* n. 137, at 172. *See* also: http://www.fao.org/docrep/W7414B/w7414b0w.htm.

[431] *See* UNGA Resolution A/RES/2248 (S-V) dated May 19, 1967, in ILM, Vol. VI, Number 1, January 1967.

[432] A question also arose as to whether notification should be undertaken by South Africa, the beneficiary of the project, or by Lesotho, where the project is located. The issue was resolved by a request to the Bank by both countries to undertake the notification on their behalf.

then the legal administering authority. This was done in November 1984, and the UN Council voiced no objection to the project in April 1985.[433] Lesotho was also asked to consult with the South West Africa People's Organization (SWAPO) which was recognized as the representative of the people of Namibia. SWAPO also had no objection to the Project. However, the Project was not presented to the Bank Executive Directors by the time Namibia gained its independence on March 21, 1990. For that reason the Bank decided to seek a reconfirmation from the newly constituted Namibian government of the no-objection received from the UN Council, and from SWAPO. Accordingly, a new notification was sent to the Namibian government in May 1990, updating the earlier notification letter and Project Details submitted to it.

Another issue raised by this Project was whether the policy applies to provinces within a federal system. In this case the question was whether the provinces of the Republic of South Africa affected by the Project would need to be notified. It was decided that application of the policy does not extend to provinces because the policy is concerned with international waterways shared by states which are subjects of international law. Provinces are not subjects of international law, and as such the policy cannot be extended to them.[434] If there are issues regarding authority over, and allocation of water resources between the provinces in a federal system, then the central government of the state concerned would need to address that internally through its own constitutional and legal framework.[435] The Bank will extend its long-standing rule of not financing a project in a waterway where a dispute exists

---

[433] In making the decision, the UN Council consulted with the Office of the Legal Affairs of the UN, and made it clear that this was without prejudice to the rights of the sovereign state of Namibia upon its independence. That was the reason the Bank decided to notify the Namibian government once it was established.

[434] A recent agreement between some provinces of Canada and some states of the United States of America (USA) raises interesting and novel issues. The Great Lakes–St. Lawrence River Basin Sustainable Water Resources Agreement was concluded in December 2005 by the states and provinces in the USA and Canada that share the Great Lakes and the St. Lawrence River, and not the governments of the USA and Canada themselves. Those States are Illinois, Indiana, Michigan, Minnesota, New York, Ohio, Pennsylvania, and Wisconsin in the USA, and the Provinces of Ontario and Quebec in Canada. The Agreement bans new diversions of water from the basin, allowing only limited exceptions, and requires the States and Provinces to use consistent standards to review proposed uses of the Great Lakes waters. It also requires that lasting economic development be balanced with sustainable water use to ensure that waters are managed responsibly. For a copy of the Agreement, *see*: http://www.cglg.org/projects/water/docs/12-13-05/Great_Lakes-St_Lawrence_River_Basin_Sustainable_Water_Resources_Agreement.pdf.

[435] This question also arose in connection with India where water under the Indian constitution is, by and large, the responsibility of the states, with a very limited role for the central government. A similar decision was made that the Bank policy does not extend to states within a federal or quasi-federal system, and that it is the responsibility of the

to cases of inter-provinces or inter-states water disputes in a federal system. In such cases, the Bank would exclude components on the river shared by the provinces that are in a dispute over that river, particularly if the dispute is pending before water tribunal in that country.[436]

The Bank also faced situations of countries in internal conflict and with no centrally recognized authority. The question that arose is who would be notified, and how should the notification process be undertaken. Rather than prescribe a rule for those kinds of situations, the Bank has decided to deal with each case and project on *ad hoc* basis. Somalia has been an example for this type of challenge. The Ethiopian Social Rehabilitation and Development Fund Project[437] initially included components on the Afar, Aysha, and Ogaden Basins which Ethiopia shares with Somalia, in addition to other components on the Blue Nile River. The issue of how to handle the notification to Somalia was deliberated within the Bank. With no central government to notify, and no accredited ambassador in Washington DC to contact, the Bank decided to drop the Project components on those basins, and limit the project to the small irrigation schemes on the Blue Nile River. Thus, no components involving rivers shared with Somalia were included under the project. However, the matter was addressed differently in two recent projects in Ethiopia.[438] Those projects included components on the Genale Dawa (which is part of the Juba Basin), Wabe Shebelle, and Ogaden Basins that Ethiopia shares with Somalia. The Bank decided to notify the Executive Director representing Somalia of the Projects and seek any comments he might have on behalf of Somalia. The change in approach from the previous project was prompted by two facts. First in 1995 it was thought and hoped that the situation in Somalia was temporary and would be resolved soon. Ten years later, the situation was still the same. The second reason was the determination by the Bank that the amount of water abstracted in the latter projects was so minimal that there would be no harm whatsoever to Somalia.

Notification to the Executive Director as the representative of the government or governments concerned was also undertaken in some projects before the 1985

---

central government to undertake the necessary arrangements for projects involving inter-states water allocation. For a discussion of allocation of responsibilities over water between the states and the central government in India, *see* Salman M. A. Salman, *Inter-states Water Disputes in India—An Analysis of the Settlement Process*, 4 Water Policy, 223 (2002).

[436] *See* India, Irrigation Project-Karnataka Tank (P009773, 1981).

[437] Ethiopia, Social Rehabilitation and Development Fund (P000771, 1996).

[438] *See* Ethiopia, Water Supply and Sanitation Project (P076735, 2004), and Ethiopia, Decentralized Infrastructure Project (2004). Processing of the latter project was discontinued in 2005 for other reasons.

OMS was issued. Since that time, the process of notification has been stream-lined. As indicated earlier, when the Bank undertakes the notification on behalf of the borrower, the notification would be sent to the Ministry of Finance of that country. When the notification is undertaken by the borrower, the Bank has in many instances sent a copy of the notification to the Executive Director of the country or countries concerned for information.[439] The Executive Directors are also involved when the borrower indicates to the Bank that it does not wish to give notification to the other riparians, and does not want the Bank to notify them on its behalf. In such cases, the Bank discontinues processing of the project and the Executive Directors concerned are informed of these developments and any further steps taken.[440]

Some projects involve all the riparians, and accordingly all those riparians would be beneficiaries of the project. These types of projects could be carried out by the river basin organization that is inclusive of all the riparians, or by a national agency of each of the riparians. Examples of such projects include "Reversing Land and Water Degradation Trends in the Niger River Basin" for which the NBA was the implementing agency, and of which all the nine riparians of the Niger Basin were beneficiaries.[441] The Environmental Protection and Sustainable Devel-opment of the Guarani Aquifer System Project involved the four riparians states of the Aquifer—Argentina, Brazil, Paraguay and Uruguay.[442] The Marine Elec-tronic Highway Demonstration Project involved the three littoral states of the Strait of Malacca, namely Indonesia, Malaysia, and Singapore, which participated in the preparation and implementation of the project.[443] In all those projects, there was obviously no need for notification because all the riparians were involved in the respective project, and were indeed the beneficiaries of the project.

The Lake Victoria Environmental Management Project II raised some inter-esting questions about which should states be notified, and by whom.[444] The proj-ect supports investments in small-scale irrigation, flood control, lake navigation aids, drainage, rehabilitation of sewerage facilities, and adoption of cleaner industrial production technologies. It also supports natural resources conserva-tion and livelihood improvement sub-projects in Kenya, Uganda and Tanzania.

---

[439] This practice was resorted to as an added measure for ensuring that the notification was indeed delivered to the right authorities in the countries concerned.

[440] *See* paragraph 4 of OP 7.50.

[441] *See supra* n. 247.

[442] *See supra* n. 241.

[443] *See* Republic of Indonesia and International Maritime Organization, Marine Electronic Highway Demonstration Project (P068133, 2006).

[444] *See* Lake Victoria Environmental Management Project II (P103298, 2008).

The Project is financed by an IDA Credit in each country. In addition, a GEF Grant is being extended to the East African Community (EAC) for financing diagnostic activities in those three countries as well as in Burundi and Rwanda, with the view that the latter two countries would join Phase II of the Project.[445] A question arose as to whether Burundi and Rwanda should be notifying or notified states for the project? It was agreed that because Burundi and Rwanda were closely involved in the preparation and appraisal of the project, and would join Phase II of the project, they would be part of the notifying states.[446] It was also noted that both countries are also members of the EAC which would carry part of the project.[447]

Despite the clarity and straightforwardness of the provisions of the policy on who could undertake the notification and who should be notified, the Bank experience has presented some complex and interesting cases. The Bank has been addressing these cases with flexibility and due diligence, thus underscoring the uniqueness of the Bank policy and its immense contribution in this field.

5.4   Content and Timing of Notification

The Bank policy requires that the international aspects of projects on international waterways are dealt with at the earliest possible opportunity. A potential international water rights issue is assessed as early as possible during project identification and described in all project documents starting with the Project Information Document (PID).[448] These aspects of the project are to be brought

---

[445] The EAC is a regional inter-governmental organization whose original members consisted of the Republics of Kenya, Tanzania and Uganda, with its headquarters in Arusha, Tanzania. The Treaty for Establishment of the East African Community was signed on November 30, 1999 and entered into force on July 7, 2000, following its ratification by the three original members. The Republic of Rwanda and the Republic of Burundi acceded to the EAC Treaty on June 18, 2007, and became full members of the EAC with effect from July 1, 2007. For more information on the EAC and the Treaty *see*: http://www.eac. int/index.php.

[446] Because Lake Victoria is one major source of the Nile River, the Bank, on behalf of the Governments of Kenya, Tanzania, Uganda, Burundi and Rwanda notified the other riparians of the Nile River, namely, the Democratic Republic of Congo, Egypt, Ethiopia, Eritrea, and Sudan, of the proposed project. There was no unfavorable response from any of them.

[447] Accession of Burundi and Rwanda to the EAC carried with it accession to the Protocol for Sustainable Development of Lake Victoria Basin, which established the Lake Victoria Basin Commission (LVBC). The Protocol was concluded by Kenya, Tanzania and Uganda on November 29, 2003. For a copy of the Protocol *see*: http://www.meac.go. ke/index.php?option=com_docman&task=doc_details&gid=30&Itemid=52. For more information on LVBC, *see infra* n. 719.

[448] *See* paragraphs 1 and 2 of BP 7.50.

promptly to the attention of senior management of the Bank, whom the Task Team are to keep regularly informed of the processing steps of the project. Concurrently, the Bank advises the prospective borrower during identification of the need to notify the other riparians of the project, and provide them with adequate information about it.

The other riparians need an adequate amount of information about the project to enable them to make an informed determination about the effects of the project on the shared waterway, and on their interests. For that purpose, the policy requires that the notification contains, to the extent available, sufficient technical specifications, information, and other data (referred to in the OP as the Project Details) to enable the other riparians to determine as accurately as possible whether the proposed project has potential for causing appreciable harm through water deprivation or pollution, or both.[449]

The policy instructs the Bank staff to ensure that the Project Details are adequate to enable the other riparians to make their determination. Such information would include a detailed description of all the components of the project, and the tentative cost of each such component to give an idea about its size. The information would include an estimate of the amount of water that would be used, compared to the overall flow of the river, or the water volume of the lake, to underscore the fact that the amount of water to be used by the proposed project is indeed minimal, and will not cause appreciable harm to the other riparians. A brief summary of the Environmental Impact Assessment (EIA) findings is also needed to show any negative effects of the project on the water quality or on the overall environment of the shared waterway and how such negative effects are avoided, minimized or mitigated.[450] Reference to agreements in force, if there are any, and how they relate to the project is usually included in the Project Details. Similarly, the institutional arrangements for the implementation of the project need to be described to allay any concerns of the other riparians about how properly the project would be implemented.

The exact location of the project is an important element of such Project Details. For that purpose, the policy requires that documentation of a project on an international waterway includes a map that clearly indicates the waterway and the location of the project's components. This requirement applies to the Project Appraisal Document (PAD), the Project Information Document (PID), and any internal memoranda that deal with the riparian issues associated with the project.[451] Maps are

---

[449] *See* paragraph 3 of BP 7.50.

[450] This issue will be discussed in more detail in Chapter 9, Part 9.3 of this Book.

[451] *See* paragraph 13 of BP 7.50. Maps are prepared and cleared in accordance with the World Bank Administrative Manual Statement (AMS) 9.50, Cartographic Services, June 2002 (formerly AMS 7.10).

included in these documentations even when notification to riparians is not required by the provisions of the policy. However, as will be discussed in Chapter 9 in connection with projects in disputed areas, the Bank has been requested by some members not to include maps of their countries in project documents because of concerns about showing certain parts of the country as disputed areas.[452] In some instances, particularly in regional projects, the Bank has published maps without demarcating the boundaries between the states involved.[453]

In some instances the Project Details and the information referred to above may not be available during the identification or preparation of the project. If adequate details are not available at the time of notification, they are made available to the riparians as soon as possible after notification. However, as will be discussed later, and as experience with notification has shown, the notified riparians would usually ask to be furnished with any information they deem necessary, if such information has not been furnished to them at the time of notification.[454] In this regard, the policy does not make an explicit mention of the need to furnish the EIA for the project to the other riparians as part of the Project Details.[455] Nonetheless, the notified riparian states are increasingly asking for copies of the EIA, particularly in projects involving dams and other water infrastructure, as will be discussed later.[456]

---

[452] The decision relates in most instances to the non-inclusion of the national map of the beneficiary state. Maps of the project area are usually included in the project documents, if no part of the project area is disputed. The decision not to include such maps is taken by the Regional Vice President in consultation with the Vice President and General Counsel.

[453] This approach has been followed in a number of instances, including maps of the Nile Basin, where there are disputes about boundaries between some of the Nile Basin states. Showing the Nile Basin without country boundaries addresses the concerns of those states, and also emphasizes the fact of the Nile as a hydrological unit.

[454] Paragraph 3 of BP 7.50 requires obtaining the approval of the Regional Vice President if it is proposed to go ahead with project appraisal before the Project Details are available. Since the Project Details will usually be available after the appraisal, it seems that the intention of the BP is to request going ahead with notification, and not appraisal.

[455] The UN Watercourses Convention requires, in Article 12, that the notification be accompanied by technical data and information, including the results of any environmental impact assessment. The Espoo Convention requires undertaking of an environmental impact assessment prior to carrying out any of the activities listed in Annex 1 to the Convention. It also requires providing an opportunity to the public, in the areas likely to be affected, to participate in relevant environmental impact assessment procedures regarding the proposed activities, and to ensure that the opportunity provided to the public of the affected Party is equivalent to that provided to the public of the Party of origin. *See* Article 2(5) of the Espoo Convention, *supra* n. 366.

[456] The Vishnugad Pipalkoti Hydro Electric Project in India is planned as a 444 MW run-of-river power plant for which a 65-meter high dam will be built. The project is situated on the Alaknanda River in the state of Uttaranchal, India. The Alaknanda River is a tributary

The other riparians are allowed a reasonable period of time from the date of dispatch of the notification letter and Project Details to respond. The policy states that the period shall normally not exceed six months.[457] GP 7.50 recommends that every effort should be made to allow the notified riparians six months to respond to the notification, and that a shorter period is given only in cases of emergency.[458] If one or more of the riparians ask for additional information or clarification, every effort would be made to provide such information, and a reasonable period would be allowed for study of the information and response. However, with the project cycle becoming increasingly short, most notification letters now ask for a response by the end of a two month period.

Most of the other international instruments discussed earlier address the issue of how much time the other riparians should be allowed for a response. The Helsinki Rules recommends ". . . a reasonable period of time to make an assessment of the probable effect of the proposed construction or installation."[459] The UN Watercourses Convention suggests, in line with the Bank policy, a similar period of six month for reply.[460] If the notified state requests an extension of the period to enable it to evaluate the information furnished and respond to the notification, the Convention suggests another period of six months. However, the Salzburg Resolution does not indicate a period for reply.

To avoid any misunderstandings with the notified states about the due date for the response, the notification letter usually includes a specific day of a specific month and year for response, rather than just request a response within two or three months. This approach has facilitated a proper processing and closure on the notification process, and has prevented disputes as to when the response should be sent.

---

of the Ganges River. The Bank notified the other riparians of the Ganges River (Bangladesh, China and Nepal) of the project. China asked for the environmental impact assessment for the project, which was not ready at the time of notification. The Bank indicated that the EIA would be furnished to China as soon as it was ready. *See supra* n. 402. Similarly, one of the components of the Bulgaria, Municipal Infrastructure Development Project (P099895, 2009) involves the completion of Plovditsi Water Supply Dam on the Iskretska stream, which is a tributary of the Arda River that flows into the Maritsa River. The Maritsa River forms part of the borders between Turkey and Greece, and finally discharges into the Aegean Sea. Both Greece and Turkey were notified of the project. Turkey asked for more information on the project, including the EIA, as well as more time to review such information. For more discussion on this issue, *see infra*, Chapter 9, Part 9.3 of this Book.

[457] *See* paragraph 4 of BP 7.50.

[458] *See* paragraph 3 of GP 7.50.

[459] *See* Article XXIX(3) of the Helsinki Rules.

[460] *See* Article 13 of the UN Watercourse Convention.

The Bank policy requires that the notification process be completed before the project is presented to the Executive Directors for consideration. It stipulates that the Project Appraisal Document (PAD) for a project on an international waterway deals with the international aspects of the project, and states that Bank staff have considered those aspects and are satisfied that: (i) the issues are covered by an agreement, (ii) the other riparians have given a positive response, or (iii) in all other cases the project will not cause appreciable harm to the other riparians.[461] The PAD would need to be finalized and distributed to the Executive Directors about a month before the project is considered by them. This in turn would actually necessitate completion of the notification process some time before that.

Furthermore, established practice of the Bank requires informing the Regional Vice President (RVP) if the notification process has not been completed by negotiations. The memo to the RVP would explain the status of notification, confirm that the project would not cause appreciable harm to any riparian, and ask for permission to proceed with negotiations pending completion of the notification process. This practice is based on the requirement to keep senior management abreast of the international aspects of the project and related events.[462]

Questions have been raised in this connection about national projects and programs, as well as projects that involve selection of sub-projects during implementation. Can notification for those sub-projects be undertaken during implementation, and how should the above requirement be complied with? The practice of the Bank in those types of projects is that staff still needs to comply with the above requirement. The project team is expected to do its utmost due diligence and work with the borrower to identify both, the international waterway that may be affected, and the possible type of effects, and complete the notification as stipulated above.[463] Allowing notification during implementation

---

[461] *See* paragraph 8 of OP 7.50.

[462] *See* paragraph 1 of BP 7.50 which reads: "A potential international water rights issue is assessed as early as possible during project identification, and described in all project documents starting with the Project Information Document (PID). The task team (TT) prepares the project concept package, including the PID, in collaboration with the Legal Vice Presidency (LEG) to convey all relevant information on international aspects of the project. When the TT sends the project concept package to the Regional Vice President (RVP), it sends a copy to the Vice President and General Counsel (LEGVP). Throughout the project cycle the Region, in consultation with LEG, keeps the Managing Director (MD) concerned abreast of the international aspects of the project and related events." In practice the MD is only involved when the RVP decides to seek advice, upon completion of the notification process, as discussed under Part 5.5 of this Chapter. Other than that, actions and decisions are being taken by the RVP.

[463] In most of these types of projects best estimates were arrived at and included in the Project Details furnished to the notified states. The Ethiopia, Social Rehabilitation and Development Fund Project (P000771, 1996) included small-scale irrigation sub-projects,

could defeat one basic rationale for notification, and that is dispute avoidance. Notification during implementation could also cause delays, especially if there was an objection. Changes in staff working on the project could also result in situations where notification may be overlooked. Another problem could arise if the borrower declines to undertake the notification itself, and would not allow the Bank to undertake notification on its behalf on one sub-project. This situation could generate more confusion and delays if it was encountered during implementation.

If the borrower made it clear from the early stages of the project that it would not notify the other riparians of sub-projects on international waterways, and would not allow the Bank to do so, then one approach followed is excluding sub-projects on international waterways. In such cases, the selection criteria for the sub-projects would exclude sub-projects on or affecting international waterways, and limit the sub-projects to those on national rivers only, as defined in the policy.[464]

Similarly, questions arose as to whether the restructuring of projects during implementation can include new components on international waterways. If the

---

which would abstract water from the Blue Nile and some of its tributaries. However, those sub-projects were demand driven and the amount of water to be abstracted could not be ascertained at the appraisal stage. The Bank undertook the notification on behalf of Ethiopia. The Project Details provided best estimates to this effect. However, some of the notified states asked for more specificity. The Bank replied that it would not be possible to provide the exact amount of water to be abstracted, and that the estimates represented the maximum amount of water to be abstracted. The concerns of those riparians were addressed by provisions in the legal documents to the effect that (i) all irrigation sub-projects involving abstractions of the waters of international waterways were subject to the IDA's prior approval; (ii) IDA would carry out annual post-reviews of water abstractions from the identified international rivers under the project; and (iii) disbursements out of the proceeds of the credit in respect of those sub-projects would cease once the amount of waters abstracted, or estimated to be abstracted, from each of the rivers reached the limits communicated to the riparian countries. *See* paragraph 3 of Schedule 1 to the Development Credit Agreement (Credit No. 2841 ET, dated May 15, 1996). Another project where the best estimates were used is Zambia, Increased Access to Electricity Services Project (P077452, 2008). The proposed Project may finance 3–5 small hydropower projects (SHP) whose capacity could range from 0.2MW to a maximum of 10MW. All of these SHPs would be run-of-river projects which would not adversely change the quality or quantity of water flows to any other riparian. The location of those SHPs would be determined during implementation, and accordingly it was not possible to ascertain the rivers where they would be located. Under those circumstances, all the riparians of the Zambezi River (Angola, Botswana, Malawi, Mozambique, Namibia, Tanzania, and Zimbabwe), and the Congo River (Angola, Burundi, Cameroon, the Central African Republic, the Democratic Republic of Congo, the Republic of Congo, Rwanda and Tanzania) were notified of the Project. The Zambezi and Congo rivers are the two main basins in Zambia.

[464] It should be remembered that national rivers flowing into semi-enclosed coastal waters are considered international waterways under the policy. *See* discussion in Chapter 4, Part 4.3 of this Book. *See* also paragraph 1(c) of OP 7.50.

new components require notification of the other riparians, then those components would be treated as a new project. Notification, accompanied by the Project Details would be furnished to the other riparian states which would be given reasonable time to respond. Implementation of the new components would have to wait until completion of the notification process, as provided for in the policy. Due to the time needed to complete the notification process, and the concerns about unexpected delays and outcomes, restructured projects have usually avoided new components on international waterways. In a few instances, the restructured project included only components dealing with rehabilitation of existing schemes which fall under the exception to the notification requirement, as discussed later.

The same approach has been extended to additional financing for on-going projects involving international waterways. If the original project involved notification of the other riparians, and the additional financing is limited to the components of which other riparians have already been notified, and there are no additional effects on the water quantity or quality of the additional financing, then there will be no need for any new or supplemental notification. However, if the additional financing would involve activities that would go beyond the original activities, or have extra effects on the quantity or quality of the flow of the water to the other riparians, then a new or supplemental notification would be required.[465]

This overall rule would also apply to projects falling under one of the exceptions to the notification requirement.[466] If the original financing falls under the exception to the notification regarding rehabilitation of existing schemes, and the additional financing is also for rehabilitation of existing but different schemes, then a new memorandum on the exception to notification, addressing the new works will be needed.[467] On the other hand, if the additional financing includes

---

[465] The Azerbaijan, Rural Investment Project, Additional Financing (P066199, 2006) included minor irrigation sub-projects in small local rivers flowing into the Caspian Sea, in addition to the activities financed under the original project (P076234, 2004). Thus, the Caspian Sea littoral states were notified by the Bank on behalf of Azerbaijan of the new activities. *See supra* n. 416.

[466] The Azerbaijan, Internally Displaced Persons Economic Development Support Project (P089751, 2008) addressed this issue. It was agreed that since the project activities under the Additional Financing were identical to those of the ongoing project, the rationale for the exception to the notification of the riparians under OP 7.50 that was obtained under the original project remains applicable, and accordingly no notification of other riparians was required for the additional financing.

[467] *See* Argentina, Infrastructure Project for the Province of Buenos Aires (APL2) (P105288, 2007) where the original and the additional financing both dealt with the rehabilitation of existing, but different schemes. *See* also Argentina, Buenos Aires Infrastructure Sustainable Investment Development Project, Additional Financing (P114081, 2009).

activities that would not fall under the exception already specified, then notification for those new activities would be required. However, thus far, additional financing provided by the Bank for on-going projects involving international waterways has funded the same components financed under the original project, with no new effects beyond those identified for the original project. The reasons for this may be cost increases in the original components, or that the new components simply did not relate to the international waterway.

A similar approach is followed with regard to adaptable program loans (APL) which usually consist of more than one phase. If the project triggers the policy, and notification is undertaken at the first phase for the entire program, then notification for the second phase would not be required as long as the notification covered the activities under this phase. The Ethiopia Productive Safety Net Project involved abstraction of water for small irrigation sub-projects from a number of international basins. The notification at the first phase covered the expected activities under the whole program. Accordingly, no notification was required for the second phase of the program.[468]

The above discussion indicates that the provisions of the policy for projects on international waterways are constantly facing new challenges and tests emanating from the new and novel types of projects that the Bank is financing. Yet, those challenges are being faced with flexible and pragmatic interpretation and application of the provisions of the policy, with due diligence and caution for the interests and rights of other riparians.

5.5 Riparians' Responses

The Bank policy specifies three possible kinds of responses that may result from the notification process: a positive response, no response, or an objection.[469] However, Bank experience has shown that the list of possible responses for notification is actually wider than that, and can include at least seven possibilities. Those possibilities are:

(i)   ***Requesting Additional Time***: One or more of the notified riparian states may ask for more time to study and evaluate the Project Details furnished to them with the notification letter. Although the Bank policy does not provide for this kind of situation, the Bank has always agreed to extend the period established for response and has set a new date, mostly corresponding with the new time frame requested by the riparian. As discussed in the previous Part of this Chapter, the UN Watercourses Convention provides for such a possibility, and

---

[468] *See* Ethiopia, Productive Safety Nets Project (APL I) (P087707, 2004), and APL II Project (P098093, 2007).

[469] *See* paragraphs 5 and 6 of BP 7.50.

recommends that a period equal to the initial period of six months be given to the notified states.[470]

(ii) ***Requesting Additional Information***: The notified riparians may claim that the Project Details furnished to them are not adequate to enable them make a determination on the effects of the project. Accordingly, they may ask for additional information. The additional information requested in many instances may relate to the quantitative effects of the project. There may be requests for the environmental impact assessment of the project which might not have been ready by the time the notification was sent to the other riparians.[471] It could also relate to some clarifications regarding the location of the project, including a request for a detailed map, if such a map was not included in the original notification. Some notified states have also requested more specificity on the amount of water to be abstracted for small-scale irrigation and water supply sub-projects in demand-driven and social funds projects.[472] In few instances, the request has gone beyond the project, and asked for more information on earlier projects that the Bank has financed on the same country and the same waterway. As such, the riparian is seeking information on the cumulative effects of Bank-financed projects there. As an international financial cooperative institution, the Bank would need to do its best to provide such information to the riparian or riparians that requested it. The request for additional information underscores the need to furnish adequate project details in the first instance so as to avoid the delays that this kind of request could entail.

One question that has arisen in connection with requests for additional information concerns the refusal of the notifying state to furnish the additional information requested by the notified state. Can the Bank provide such information to the notified state even if the notification was undertaken by the borrowing riparian state? As a first step, the Bank would discuss the matter with the notifying state, and ascertain the reasons for its refusal to provide such information itself. The Bank would try to persuade that state to furnish the requested information if it found the request for such information is reasonable and warranted. If the state persists in its refusal, the Bank could offer to furnish such information on its behalf. If the state refuses to allow the Bank to furnish such requested information, and the Bank ascertained that the need for such information is reasonable, then the Bank would extend application of paragraph 4 of the OP 7.50 which requires the Bank to discontinue processing of the project if the borrower objects to giving notification and objects to the Bank doing so.

---

[470] *See* Article 13 of the UN Watercourse Convention.

[471] *See* Vishnugad Pipalkoti Hydro Electric Project *supra* n. 402.

[472] This request was made by Egypt and Sudan in connection with the Ethiopia, Social Rehabilitation and Development Fund Project, *see supra* n. 437.

Another issue discussed repeatedly is how the Bank should respond to requests for additional information received by the Bank after the date set for response in the notification letter. This is dealt with in (iii) below.

(iii) *Requesting Additional Information and Time*: In this case, the request would combine (i) and (ii) above. The notified riparian would state that it needs additional information, and additional time, to evaluate this information. The request for additional information could identify the specific information needed, or be in the form of questions about the proposed project. The request could be sent immediately after the notification letter was sent. It could also be sent towards the end of the period provided for sending a response.[473] This situation is provided for in GP 7.50 which recommends providing both the requested information, and the additional time needed to evaluate it. Requests for additional information and time together are more common than either (i) or (ii) above.

Requests for additional information and additional time received by the Bank from the notified state after the date set for response in the notification letter are dealt with on a case-by-case basis, depending on the timing and type of information requested.[474] If the request is received a few days after the date set for response, and the information requested may be relevant to the decision making, and there is still time for presentation of the project to the Executive Directors, then the Bank has provided the requested information and set a new date for response. However, if the request is received some time after the date set for response, or the information requested is not that important to making the decision, then the Bank would send the information, reconfirm that the project would not cause appreciable harm, and explain that it would proceed with project processing as planned.[475] The Bank would also encourage the borrower, if it has undertaken the notification itself, to follow those same procedures. If the request for information is received after the project was approved by the Executive Directors, then the information would be provided and this situation stated in the reply to preclude further requests for similar information.

(iv) *No Response*: In this situation, no response would be received by the borrower or the Bank (in case the Bank has undertaken the notification on behalf of the borrower) from all, or some of the riparians by the date set in the notification letter for such a response. Before making a determination that the other riparian states have not responded by the date specified in the notification letter, the

---

[473] *See* Bulgaria, Municipal Infrastructure Development Project, *supra* n. 415.

[474] *See* paragraph 4 of GP 7.50.

[475] Those decisions would be made by senior management of the Bank on the recommendation of the Task Team, following the same procedures provided for under paragraph 4 of BP 7.50.

Bank usually ensures that the notification letter was indeed received by the other riparian states. This is particularly important if the notification is being undertaken by the borrower and not the Bank. In such cases, the Bank needs to satisfy itself first that the notification was both sent, and also received by the right ministry of the states that are being notified. One way of ensuring that this has happened is to send a copy of the notification sent by the borrower to the Executive Directors representing the notified riparian states immediately after the notification is sent to the other riparians.

The absence of response from the notified states has been one main feature of the notification process for the Bank projects. Some states do not want to give an explicit consent or objection to the project, so they simply do not respond. The question that arises is whether failure to respond would be construed as a no-objection to the project. The Bank has taken that position based on the determination that the projects it finances do not cause appreciable harm to other riparian states. Failure of the notified state to challenge that determination, when they have been given the opportunity to do so, is taken by the Bank as an implicit confirmation that the project will indeed not cause appreciable harm to the notified states. The Bank has usually described this situation as "absence of unfavorable response." It is worth noting that the UN Watercourses Convention allows the notifying state to proceed with implementation of the planned measures, subject to certain conditions.[476]

(v)   *Positive Response or No-objection*: The positive response from some or all the riparians could take one of several forms. It could be a consent; or even an enthusiastic support for the project; it could be a confirmation of the determination made by the Bank that the project will not harm their interests; or it could simply be a general no-objection to the project.[477]

---

[476] Article 16 of the Convention states that "1. If, within the period applicable pursuant to article 13, the notifying State receives no communication under article 15, it may, subject to its obligations under articles 5 and 7, proceed with the implementation of the planned measures, in accordance with the notification and any other data and information provided to the notified States. 2. Any claim to compensation by a notified State which has failed to reply within the period applicable pursuant to article 13 may be offset by the costs incurred by the notifying State for action undertaken after the expiration of the time for a reply which would not have been undertaken if the notified State had objected within that period."

[477] It is worth noting that the UN Watercourses Convention does not explicitly address the possibility of a positive response. The Convention only indicates that the notified states would communicate their findings to the notifying state as early as possible within the specified period. It goes on to address the situation where the notified state finds the implementation of the planned measure to be inconsistent with Articles 5 or 7 of the Convention (equitable and reasonable utilization, and the obligation not to cause harm, respectively).

Despite the difficulties surrounding international waterways by reason of the competing demands of the different riparians, there have been a good number of cases where riparians have provided some form of no-objection to Bank-financed projects. This positive response is usually facilitated by the confirmation by the Bank, when it sends the notification, that the project will not cause appreciable harm to any of the riparians, a determination that should obviously be supported by the data and information laid out in the Project Details. The positive response can also be received from the river basin organization on behalf of the riparian states when the notification was sent to such organization.

(vi) *Qualified No-objection*: There have been instances where the borrower or the Bank has received a qualified no-objection from one of the riparian states. Such a response would indicate that the state concerned has no-objection to the project, provided that certain actions are taken. In one project involving the abstraction of water from a shared aquifer, the notified state gave its no-objection to the project provided that no drilling of wells would take place within a certain distance of the borders of each of the two countries.[478] This was accepted by both the Bank and the borrower because of the reciprocal nature of the condition. In another project, one of the notified states gave its no-objection to the project provided that the Bank and the borrower would supply that state regularly during the project period with reports on the implementation of the environment management plan for the project and the actual effects on the project on that riparian.[479] In its rejection of this condition, the Bank reiterated its determination that the project would not cause appreciable harm to the other riparians, and clarified that periodic reports are usually issued for all Bank-financed projects which can be accessed and reviewed by any interested party.[480]

There may be situations where it may not be easy to distinguish between requests for additional information and the case of a qualified no-objection. This determination would depend on the wording of the response received from such a notified state. The notified state may ask for a cumulative impact assessment of all Bank-financed projects on that specific waterway, to ascertain the effects of all these projects, and not just that particular project. This could be a request for additional information of such a cumulative impact statement. It could also be a condition for its no-objection. In one such case the Bank confirmed the *de minimis* effects of the project at hand, and clarified that a cumulative impact assessment

---

[478] *See* Jordan, Disi-Amman Water Conveyor Project (P051749, 2001), *infra* n. 642.

[479] *See* Nepal, Power Development Project, *supra* n. 402.

[480] The World Bank Project Status web page includes information on each operation funded by the Bank. This information is available at: http://go.worldbank.org/PGQUQ521H0.

was beyond the scope of that project, but could be done as part of the EIA in one future project. Thus, the Bank classified the request as one for additional information which was not available, rather than a qualified no-objection.[481]

Most of the above demands by the notified states could be classified as qualified no-objection. However, there may be cases where the qualification to the no-objection is more in the nature of a conditional no-objection, than a qualified no-objection. Such a determination would be arrived at depending on the nature of the response. For example if the notified state questioned the height of the dam to be constructed under the project, or the design of that dam, and asked for decreasing the height or changing the design, then this would most likely be classified as a conditional no-objection. Under such a situation, the Bank and the borrower would provide clarifications to the notified state as to why the project should be kept as proposed. If the notified state insists on its condition, then this conditional no-objection would be treated as an objection and be dealt with under the procedures set forth below.

(vii) *Objection by One or More of the Riparians*: The Bank is required to confirm in the notification letter and the Project Details that the project will not cause appreciable harm to any of the riparians, or will not affect the quantity or quality of the waters of the shared waterway.[482] Still both the determination and confirmation have been challenged by some riparians in some projects, as we have seen in the previous Chapters of this Book. The examples of the Ghab Project, and the Igdir-Aksu Project were discussed in details earlier. A detailed discussion of the issue of objection by some riparians to Bank-financed projects on international waterways is given in the next Chapter of this Book.

One question that has arisen in connection with objections is whether the Bank should follow those procedures outlined below if the objection is received from the notified state after the date set for response in the notification letter. Again, the issue has been handled on a case-by-case basis, depending on the timing of the objection, and the reasons for it.

If the objection is received a few days after the date set for response, and there is still time for presentation of the project to the Executive Directors, then the Bank would follow the same procedures set forth below. If the objection is

---

[481] The Ethiopia, Urban Water Supply and Sanitation Project (P101473, 2007) involved abstraction of water from the Omo River, which flows into the Lake Turkana. The Lake is shared by Ethiopia and Kenya, with the largest part of the Lake falling within Kenya. Kenya asked for more information on the service level and net abstraction, as well as the effects of other projects on the Omo River. The Bank confirmed the minimal effects of the project, and indicated that it would make the EIA available once it was completed.

[482] As discussed earlier, in the presentation of the project to the Executive Directors of the Bank, the PAD needs to confirm that the project would not cause appreciable harm to other riparians. *See* paragraph 8 of OP 7.50.

received after the project documents have been distributed to the Executive Directors, then the Task Team would seek the guidance of senior management. However, if the request is received after the project has been approved by the Executive Directors, then the case for following the procedures is already a moot one. In such a case a reply would be sent to the objecting state reconfirming that the project will not cause appreciable harm to that state, and explaining the stage that project processing has reached.

The above brief explanation of the actual working of the policy has shown that the responses of the riparians have varied, and taken forms not anticipated at the time of preparing the policy in 1985. Yet, the Bank, as an international financial cooperative institution, has addressed each of these novel situations with utmost care, keeping in mind the interests of both the borrower and the other members of the institution.

CHAPTER **6**

Objections to Bank-Financed Projects

6.1   Bank Procedures for Dealing with Objections

As we have seen in the previous Chapters of this Book, the objection to the Ghab Project in Syria, and to the Igdir-Aksu-Eregli-Ercis Irrigation Project in Turkey, both had played a significant role in the evolution of the Bank policy for projects on international waterways. Turkey's objection to the Ghab Project prompted the Bank to issue its first policy in 1956, establishing an early warning system for a close, careful and cautious look at those types of projects. As discussed earlier, in 1965 OM 5.05 added some substantive and procedural rules to OM 8. It took 20 years and another objection for the Bank to issue another more detailed and elaborate policy in 1985, replacing its 1965 OM. That objection came from Iran over a project in Turkey. Since that time a number of ambiguities and gaps in the policy were filled through the implementation experience, including how to deal with the different kinds of responses.

The Bank has put in place detailed procedures for dealing with responses. In the case of a positive response, or absence of response, the Country Director, in consultation with the Legal Vice Presidency and other departments concerned, addresses a memorandum to the Regional Vice President (RVP). The memorandum reports all relevant facts, including the staff assessment of whether the project would (a) cause appreciable harm to the interests of the other riparians, or (b) be appreciably harmed by the other riparians' possible water use. The memorandum seeks approval for further action. In making this decision, the RVP seeks the advice of the Managing Director (MD) concerned.[483] Two observations can be made about these procedures.

First, the procedures equate the positive response with the absence of response. If there is a positive response, there is really no need for seeking approval for further action since the positive response would remove any obstacles to further processing of the project. Actually, in practice, the procedures are only followed in case of an absence of response.

Second, the procedures injected the Managing Director in the decision-making process. The reason for this is an attempt to add an objective view to the process. The project would usually be processed by the region, and if the decision

---

[483] *See* paragraph 5 of BP 7.50.

is left entirely to the Regional Vice President, then this process may not be seen as objective. Moreover, the role of the Managing Director would bring a Bank-wide perspective to the issue of absence of response, and ensure consistency and objectivity in addressing the issue.

In case of an objection, the procedures set forth by the Bank are understandably more detailed, and are spelled out in paragraphs 6–12 of BP 7.50.[484] The policy requires the Country Director (CD), in collaboration with the Legal Vice Presidency (LEGVP) and other departments concerned, to send a memorandum on the objection to the RVP, with a copy to the LEGVP. The memorandum addresses:

(a) the nature of the riparian issues;
(b) the Bank staff's assessment of the objection raised, including the reasons for such an objection, and any available supporting data;
(c) the staff's assessment of whether the proposed project will cause appreciable harm to the interests of the other riparians, or be appreciably harmed by the other riparians' possible water use;
(d) the question of whether the circumstances of the case require that the Bank, before taking any further action, urge the parties to resolve the issues through amicable means such as consultations, negotiations, and good offices[485] (which will normally be resorted to when the other riparians' objections are substantiated); and
(e) the question of whether the objections are of such a nature that it is advisable to obtain an additional opinion from independent experts.

The RVP seeks the advice of the MD concerned, and the LEGVP, and decides whether and how to proceed. On the basis of these consultations, the RVP may recommend to the MD concerned that the Operations Committee consider the matter.[486] The CD then acts upon either the Operations Committee's instructions,

---

[484] *See id.*, paragraphs 6–12.

[485] The approaches of negotiations and good offices under the policy are merely suggestions to the parties and not a requirement. Paragraph 3 of the OP states that in case the differences remain unresolved between the beneficiary state and the other riparians, prior to financing the project, the Bank normally urges the beneficiary state to offer to negotiate in good faith with the other riparians to reach appropriate agreements or arrangements. Paragraph 6(c) of BP 7.50 referred to above, also suggests that the Bank could "urge the parties to resolve the issues through amicable means, such as consultations, negotiations, and good offices."

[486] The Operations Committee replaced the Loan Committee in 1996 (*see supra* n. 99), and the Committee's mandate was changed to concentrate on the review of Country Assistance Strategies and on select operations which warranted Bank-wide attention because of their policy implications, risks, or their innovative nature. *See*: http://web.worldbank.org/WBSITE/EXTERNAL/EXTABOUTUS/EXTARCHIVES/0,, content-MDK:20677038~pagePK:36726~piPK:36092~theSitePK:29506,00.html.

which are issued by the chairman, or the RVP's instructions, and reports the outcome in a memorandum prepared in collaboration with LEGVP and other departments concerned. The memorandum, sent to the RVP and copied to the LEGVP, includes recommendations for further processing of the project.

If a decision is taken to seek the opinion of independent experts, before further processing of the project, the RVP requests the Vice President, Sustainable Development Network (SDNVP) to initiate the process. The SDNVP maintains, in consultation with the RVPs and LEGVP, a roster of highly qualified independent experts, which consists of 10 names, and is updated at the beginning of each fiscal year. The SDNVP, in consultation with the RVP and LEG, selects one or more independent experts from a roster maintained by SDNVP. The experts selected may not be nationals of any of the riparians of the waterways in question, and also may not have any other conflicts of interest in the matter.[487] The experts are engaged and their terms of reference prepared jointly by the offices of the SDNVP and the RVP. The latter finances the costs associated with engaging the experts. The experts are provided with the background information and assistance needed to complete their work efficiently.

The experts' terms of reference require that they examine the Project Details. If they deem it necessary to verify the Project Details or take any related action, the Bank makes its best efforts to assist. The experts meet on an *ad hoc* basis until they submit their report to the SDNVP and the RVP. The SDNVP or RVP may ask them to explain or clarify any aspect of their report.[488] It is worth emphasizing that the experts have no decision-making role in the project's processing. Their technical opinion is submitted for the Bank's purposes only, and does not in any way determine the rights and obligations of the riparians.[489] Their conclusions are reviewed by the RVP and SDNVP, in consultation with the LEGVP.

A number of observations can be made about this process:

First, the objecting riparian has no veto power over the project, and its objection does not prevent the Bank from financing the project. In fact no international instrument grants the objecting state such veto power. The requirement for the consent of other riparians under the Madrid Declaration of 1911 was replaced by

---

[487] *See* paragraph 9 of BP 7.50. It is worth noting that this paragraph requires that the experts selected may not be nationals of any of the riparians of the waterways in question, and not just of the notifying and objecting states.

[488] *See id.*, paragraph 10.

[489] *See id.*, paragraph 11. It is noteworthy that Annex A to OMS 2.32, as well as Annex A to OD 7.50, were both entitled "Technical Advice of Independent Experts." However, paragraphs 8–11 are together entitled "Seeking the Opinion of Independent Experts." The change from "Advice" to "Opinion" is, in my view, intended to emphasize further that the experts have no decision-making role.

the requirement of consultation and negotiations by the Salzburg Resolution of 1961. In fact the Lake Lanoux arbitration had declared a few years before the Salzburg Resolution that consent of other riparians to the proposed project is neither a custom, nor a general principle of law.[490]

Second, before starting the independent experts' process, the policy suggests urging the parties to resolve the issues through amicable means such as consultations, negotiations, and good offices. This route will normally be resorted to when the other riparians' objections are substantiated because the Bank will not be able to finance the project if it is established that it will cause appreciable harm without the consent of the affected riparian or riparians.

Third, the process of the independent experts involves a number of units within the Bank: the Region where the project is being processed, the Legal Vice Presidency which oversees the policy for projects on international waterways, and the Sustainable Development Network that deals with environment and natural resources. The latter is even required to maintain a roster of names for such experts. The involvement of those three units precludes conflict of interest situations, gives the process credibility and objectivity, and ensures its fairness.

Fourth, the issue of the possibility of conflict of interest is adequately addressed. The experts selected may not be nationals of any of the riparians of the waterways in question, and not just of the notifying and notified states. The experts may not have any other conflicts of interest in the matter.

Fifth, the process is not an arbitration with binding decisions. On the contrary, the experts have no decision-making role in the project processing. Their technical opinion is submitted to the Bank, and the Bank decides what to do with it. Furthermore, their opinion does not in any way determine the rights and obligations of the riparians. Nonetheless, the Bank may ask them to explain or clarify any aspect of their report.[491]

These procedures contrast sharply with the procedures set forth by the IIL, ILA and the UN Watercourses Convention, as discussed below.

6.2   Procedures under other International Instruments

The IIL, ILA and ILC each adopted provisions addressing the issue of how to deal with an objection by one of the riparian states to a proposed project. The Salzburg

---

[490] The Arbitral Tribunal stated that "The rule according to which States may utilize the hydraulic force of international watercourses only on condition of a *prior* agreement between the interested states cannot be established as a custom, or even less as a general principle of law." *See supra* nn. 82 & 356; 24 ILR 101 (1957) at 130.

[491] *See* paragraph 10 of BP 7.50.

Resolution issued by the IIL in 1961[492] requires, in case an objection is made to a project on a shared watercourse, the states to enter into negotiations with a view to reaching an agreement within a reasonable time. The Resolution recommends that those states have recourse to technical experts and, should occasion arise, to commissions and appropriate agencies in order to arrive at solutions assuring the greatest advantage to all concerned. It goes on to oblige the notifying state, in conformity with the principle of good faith, to refrain from undertaking the works or utilization which are the object of the dispute, or from taking any other measures which might aggravate the dispute, or render agreement more difficult. If the interested states fail to reach agreement within a reasonable time, the Resolution recommends that the parties submit to judicial settlement or arbitration the question of whether the project is contrary to the above rules.[493]

The ILA Helsinki Rules of 1966 do not deal with an objection to a proposed project *per se,* but rather with a question or dispute relating to the present or future utilization of the waters of an international drainage basin. The Rules recommend that the basin states try to resolve such a question or dispute through negotiations.[494] Failing that, they may refer the question or dispute to a joint agency and that they request the agency to survey the international drainage basin and to formulate plans or recommendations for the fullest and most efficient use thereof in interests of all such States. If the question or a dispute cannot be resolved in this manner, the Rules recommend that the parties seek the good offices, or jointly request the mediation of a qualified international organization or of a qualified person. If the parties fail to resolve their dispute through those procedures, the Rules recommend that they form a commission of inquiry or an *ad hoc* conciliation commission, which shall endeavor to find a solution.[495] Finally, the Rules recommend submitting the dispute to an *ad hoc* arbitral tribunal or to the International Court of Justice if the commission could not reach a solution, or the solution was not accepted by the parties.

---

[492] *See supra* n. 69.

[493] *See* Articles 6–8 of the Salzburg Resolution, *supra* n. 69. According to Article 8, if the state objecting to the works or utilization refuses to submit to judicial settlement or arbitration, the other state is free, subject to its responsibilities, to go ahead while remaining bound by its obligation arising from the provisions of Articles 2–4 (the rights of utilization by other states, equitable sharing and the right to compensation in case of losses).

[494] The Helsinki Rules devote the entire Chapter 6 for dispute prevention and dispute settlement. Articles XXX to XXXVII of Chapter 6 deal with dispute settlement.

[495] The Helsinki Rules recommend that the conciliation commission be constituted in the manner set forth in the Annex to the Helsinki Rules entitled "Model Rules for the Constitution of the Conciliation Commission for the Settlement of a Dispute." *See supra* n. 166.

The procedures proposed under the UN Watercourses Convention are, by and large, similar to the procedures set forth by the ILL and the ILA. The UN Watercourses Convention states that if a notified state finds that implementation of the planned measures would be inconsistent with Articles 5 and 7 of the Convention (regarding equitable and reasonable utilization and the obligation not to cause significant harm, respectively), it shall attach to its finding a documented explanation setting forth the reasons for the finding.[496] Subsequently, the notifying state and the notified state shall enter into consultations and, if necessary, negotiations with a view to arriving at an equitable resolution of the situation. The consultations and negotiations shall be conducted on the basis that each state must in good faith pay reasonable regard to the rights and legitimate interests of the other state.[497] During the course of the consultations and negotiations, the notifying state shall, if so requested by the notified state at the time it makes the communication, refrain from implementing or permitting the implementation of the planned measures for a period of six months unless otherwise agreed.[498]

In addition to these special provisions dealing with consultations and negotiations concerning planned measures, the UN Watercourses Convention includes a separate article on the settlement of disputes.[499] This Article lays out a number of procedures for resolving disputes, including fact finding,[500] as well as the options of submitting the dispute to the ICJ, or to arbitration.[501]

The sharp contrast between the IIL, ILA and the UN Watercourses Convention procedures on the one hand, and those of the Bank on the other hand, stems from the basic difference regarding the nature of the projects in each case. The projects or planned measures would seriously or substantially affect other riparians in the case of the Salzburg Resolution and the Helsinki Rules, respectively. They could have significant adverse effects in the case of the UN Watercourses

---

[496] *See* Article 15 of the UN Watercourses Convention, *supra* n. 180.

[497] *See id.*, Article 17.

[498] Article 19 of the UN Watercourses Convention (*see id.*) states that in the event that the implementation of planned measures is of the utmost urgency in order to protect public health, public safety or other equally important interests, the state planning the measures may, subject to Articles 5 and 7, immediately proceed with implementation. In such case, a formal declaration of the urgency of the measures shall be communicated without delay to the other watercourse states. However, the states planning such measures shall at the request of any of the other watercourse states enter into consultations and negotiations with such a state.

[499] *See* Article 33 of the UN Watercourses Convention.

[500] The Article asks the parties to consider the report of the fact finding commission in good faith.

[501] The UN Watercourses Convention includes a detailed Annex on Arbitration consisting of 14 articles.

Convention.[502] However, as explained earlier, the Bank will not finance a project that would cause appreciable harm to other riparians. If it is established that appreciable harm could be caused by the project to other riparians, then the Bank will either drop the entire project, or the component that would cause appreciable harm. If the objection to the project is substantiated, the policy suggests urging the parties to resolve the issues through amicable means such as consultations, negotiations, and good offices. The Bank will only finance the project that may cause appreciable harm if the issues are resolved through negotiations, and an agreement is reached between the riparians on the project. The independent expert process is meant to confirm that the project will indeed not cause appreciable harm to any of the riparians, and is resorted to when the objection of the notified riparian or riparians is substantiated.

Accordingly, the procedures established by the Bank are far less complex than those proposed by the ILA or ILC. The purpose behind the Bank procedures is simply to ascertain if any harm could be caused by the project. That was the essence of the opinion of the independent experts which was sought and obtained in one Bank-financed project, as discussed below.

6.3   Seeking the Opinion of Independent Experts

The Bank has received very few objections for the projects it has been financing on international waterways. The main reason for this situation is that the Bank ensures from the very beginning that its financed projects will not cause appreciable harm to any riparian. The purpose of the notification is basically to convey and confirm this fact about its project to the other riparians, and obtain a reconfirmation to this effect from them. The notification letter and the accompanying Project Details often provide adequate information to this effect. In some instances, the notified riparians may ask for more information about the project, and the Bank ensures that such information is provided to them, with an emphasis that the project will not cause appreciable harm. This approach has resulted in very few objections to Bank-financed projects on international waterways.

Furthermore, most of those objections stem from the poor relations between some of the riparian states themselves, rather than from matters related to the project. Issues not pertinent to the project such as the absence of notification by the beneficiary state of other past projects, or failure of the borrower to become a party to a regional agreement on international watercourse, or to the river basin in question, have been raised by some of the objecting riparians. The presence of a border dispute has also been a reason for objection to projects on the shared

---

[502] *See supra* n. 371 for clarification by the Working Group, as part of the Statements of Understanding, of what is meant by "significant adverse effects."

waterway.[503] The Bank has decided that the objections to its financed projects, except the one described below, did not have sufficient merit to warrant seeking the opinion of independent experts. Accordingly, the Bank proceeded with the processing of these projects without appointing independent experts to look into the objections.

One such a project was the Karakaya Hydropower Project in Turkey.[504] The project consisted of the construction of a hydropower plant with a reservoir and a 173 meter-high dam on the Euphrates River which Turkey shares with Syria and Iraq. Both Syria and Iraq were notified of the project and provided with the Project Details. As part of the preparation for the project, the "Rule of 500 was developed" according to which Turkey would maintain a flow of no less than 500 cubic meters per second in the Euphrates at the Turkey-Syria borders during the filling of the Karakaya reservoir. Iraq objected to the project stating that it would not agree to any project on the Euphrates River prior to the conclusion of an agreement on the equitable sharing of the waters of the river by the three riparians. Syria also objected stating its dissatisfaction with the "Rule of 500" claiming that this volume of water represented only over half the flow of the Euphrates, and demanded a more permanent and equitable arrangement. The Bank assessed the objections from the two riparians and concluded that (i) it could rely on the formal representation of Turkey regarding the Rule of 500, which adequately protected the interests of Syria and Iraq, (ii) the consumptive use of the project would only last while the reservoir was being filled, and during that period the Rule of 500 would prevail,[505] and (iii) the Bank could only offer assistance for an agreement if all three riparians requested it; and moreover this matter was beyond the scope of the proposed project. Based on the above, the Bank decided to proceed with the project processing despite the objections by Iraq and Syria. The Loan Agreement for the project included a reference to the Rule of 500.[506] For these

---

[503] This has been particularly true with regard to notification between Eritrea and Ethiopia, as well as Armenia and Azerbaijan.

[504] Turkey, Karakaya Hydropower Project (P008930, 1980).

[505] For more on the Karakaya Project and the Rule of 500, *see* John Kolars, *Problems of International River Management, The Case of the Euphrates*, in *International Waters of the Middle East—From Euphrates to Nile*, at 44 (Asit Biswas, ed., Oxford University Press 1994).

[506] Preamble (B) to the Loan Agreement for the Project between the International Bank for Reconstruction and Development and Republic of Turkey stated that "In order to safeguard adequately the interests of the lower riparian states, the Borrower will implement certain measures with respect to the construction, filling and operation of the dam included in the Project as set forth in the letters dated March 7, 1979, and February 22, 1980, addressed to the Bank by the Borrower's Minister of Finance." Those letters referred to the Rule of 500.

reasons the idea of appointing independent experts to look into those objections was not considered by the Bank.

Objection by one riparian state to the project when some of the other riparian states have responded positively, or have not responded at all, usually indicates the weak basis for the objection. As discussed earlier, in a number of instances such objection would be prompted by factors beyond the proposed project. Furthermore, when the relationship between two neighbors are not friendly, objections to projects on shared waterways are to be expected, but such objections are usually based on the general unfriendly relations, rather than on any adverse effects by the proposed project on the shared waterway and that particular riparian. More importantly, the Bank usually exerts an extra effort to ensure that the project would not cause appreciable harm to any of the other riparians. For those reasons, appointment of independent experts to look into an objection by one of the riparians has been a very rare occasion.

The process of seeking the opinion of independent experts was put into effect in the Baardhere Dam Project in Somalia.[507] Indeed, this is the only project in the history of the policy where such experts were appointed. The project consisted of the construction of a multipurpose dam about 600 meters long and 75 meters high, to regulate the flow of the Juba River and to generate hydropower. The project also included the construction of irrigation and drainage systems for about 5,000 hectares of land. The proposed Baardhere Dam was located on the Juba River, 35 kilometers upstream of the town Baardhere in Somalia. The Juba River is fed by three main tributaries: the Genale, the Wabe Gestro, and the Dawa which originate in Ethiopia, with the Dawa becoming a boundary river with Kenya. Those rivers converge at Dolo, near the Somali borders with Kenya and Ethiopia, forming the main Juba River. The River then flows through Somalia before emptying into the Indian Ocean.[508] Thus, Ethiopia is the upstream riparian and Somalia is the downstream riparian, while Kenya is a riparian by virtue of the Dawa tributary. No agreement has been concluded on the river, nor have any basin management arrangements been established.

Processing of the project started in 1983. In 1985, following adoption by the Bank of OMS 2.32 "Projects on International Waterways," the Bank informed Somalia of the need to notify Ethiopia of the project. However, because of the strained diplomatic relations between the two countries at that time, Somalia asked the Bank to undertake the notification on its behalf.

---

[507] Somalia, Baardhere Dam Project (P002492, 1983).

[508] For more details about the Juba River, *see* Dante Caponera, *National and International Water Law and Administration—Selected Writings*, at 221–222 (Kluwer Law International 2003).

Consequently, the Bank notified the Government of Ethiopia of the proposed Baardhere Dam Project in September 1986. It provided information on the project's basic design features to enable Ethiopia to make its own determination as to the likely impact of the project, and stated that in the Bank judgment, the project would not cause appreciable harm to Ethiopia. Pursuant to the Bank policy, Ethiopia was asked to send any comments it might have within a period of six months, hence no later than March 1987.

In its reply of March 1987, the Ethiopian Government took the position that Ethiopia had the potential to impound all the discharge of the Juba River for irrigation and hydropower development, and that the project did not take into consideration present and future irrigation and hydropower developments within Ethiopia. Thus, Ethiopia's response also conveyed the concerns that the project would foreclose the potential for its future uses of the waters of the Juba River.[509] The Government of Ethiopia suggested prior negotiation with Somalia concerning the use of the waters of the Juba River, with the view of reaching an agreement which would be mutually satisfactory to both riparian states. Ethiopia went further and suggested negotiations with Somalia to determine the amount of water of the Juba River that each country would be allocated.

The Bank concluded that Ethiopia's reaction to the notification amounted to an objection to the proposed project. The Bank was also concerned that the future uses of Ethiopia could appreciably harm the project. As a result, the Bank proposed that Somalia should hold negotiations with Ethiopia over the Juba River. At that stage, Kenya's role as a riparian was noted for the first time. Consequently, the Government of Somalia was advised of the need to notify Kenya of the proposed project. In response, Somalia asked the Bank to notify Kenya on its behalf, and this was done in November 1987. Kenya did not respond to the notification.

The Somali Government was not amenable to the idea of negotiations with Ethiopia, or the use of good offices of a third party to facilitate a resolution of the issues raised by Ethiopia. Instead, Somalia proposed that the Bank should make its own assessment of Ethiopia's objection and claims. Ethiopia's proposal of negotiations conferred credibility on its objection, and made it difficult for the Bank to ignore it. Under these circumstances the question arose whether— for the first time—the matter should be referred to independent expert pursuant

---

[509] As indicated earlier, Ethiopia, the upstream riparian of the Nile River, protested against the Toshka project, *see supra*, n. 391. Ethiopia's objection to the Baardhere Dam Project reflected its concern that those waters would not be available in the future when Ethiopia would become ready to use them. Thus, Ethiopia, the upstream riparian in both instances, claimed that the other downstream riparians can cause harm to it. These positions are consistent and reflect Ethiopia's comprehension of the concept of foreclosure of future uses.

to paragraphs 11(e) and 13 of OMS 2.32 of 1985, and its Annex A, "Technical Advice of Independent Experts."

After extensive internal deliberations, the Bank decided to proceed with the appointment of independent experts. Three such experts were duly appointed in October 1988, after thorough and extensive preparations.[510] The terms of reference required the independent experts to: (a) examine the nature of the riparian issues; (b) assess whether the proposed project would cause appreciable harm to other riparians or would be harmed by the use of water by other riparians; and (c) give an opinion on the staff assessment of the riparian issues. Specifically, the experts would examine the Project Details and other studies, and as necessary take steps to verify Ethiopia's claims on past and proposed uses of the Juba River and its tributaries. The experts would also assess the viability of the project if water flow in the proposed Baardhere Dam is depleted, reduced or altered by Ethiopia.[511] To ensure their independence, the experts were appointed and expected to report to the Senior Vice President, Operations; and not to the Vice President of the Africa Region where the project was being processed.[512] Their terms of reference included the possibility of field visits to Somalia, Ethiopia and Kenya.[513]

The independent experts met during the last week of May 1989, and submitted their report on May 31, 1989.[514] The report discussed both the technical issues associated with the project, as well as the basic principles of international water law.[515] The independent experts stated that the Baardhere dam would cause no appreciable harm to Ethiopia or Kenya as the dam was downstream of both states,

---

[510] The independent experts selected were Mr. Lloyd A. Duscha, Dr. Boonrod Binson (both engineers), and Dr. Dante Caponera (a water lawyer). Mr. Duscha was selected by the other two experts as the Chair of the group.

[511] Some discussion took place on whether the Bank should consult Ethiopia and Somalia on the terms of reference of the independent experts. However, it was decided that those were the Bank's independent experts, and hence there was no reason for the Bank to consult the parties.

[512] Paragraph 2 of the Annex to OMS 2.32 of 1985 "Technical Advice of Independent Experts" stated that the Vice President, Operations Policy Staff "will, when so requested, select the experts from the roster in consultation with the Regional Office concerned and the Legal Department, to ensure the appropriate expertise and terms of reference. The experts should not be nationals of any of the riparians of the river or of other waterways involved and should have no conflict of interest in the matter."

[513] Paragraph 10 of BP 7.50 states that if the experts "deem it necessary to verify the Project Details or take any related action, the Bank makes its best efforts to assist."

[514] *Report of Independent Experts on Riparian Issues*—Somalia: Baardhere Dam Project, May 23–31, 1989, World Bank Headquarters Washington, D.C., Submitted by Lloyd A. Duscha, Chairman, Dr. Boonrod Binson, Dr. Dante Caponera.

[515] The independent experts cited a number of conventions and treaties of general relevance to the three riparian countries. They also cited the only treaty directly related to the

and its reservoir limit was well within Somalia. It accepted as valid the analysis of the Bank staff that projected a 20 to 36 percent decrease in mean annual run-off due to abstractions and evaporations in Ethiopia.[516] The experts also noted that it was possible to decrease of the flow of the Juba River upstream of Somalia to the severe detriment of the project. Equally important was the observation of the experts that the project could become a detriment to Ethiopia by virtue of prescriptive rights gained with time.[517] This point was mentioned again in connection with the general discussion of the principles of international water law. In that connection the experts observed that "present reasonable uses shall take into consideration the legitimate future requirements of a co-basin state which at present is not in a position to develop the same water resources in its territory."[518] The experts also underscored the principle of international water law of equitable and reasonable utilization of the waters of the Juba River among the three riparians, and thought that it was indispensible that negotiations between those riparians should take place with a view of concluding a framework agreement.[519]

The independent experts concluded that, considering the data available to it, the Bank staff had performed a reasonable and prudent analysis in developing its assessment, and that without direct knowledge of the actual situation or the future plans of Ethiopia particularly, the independent experts were in no position to dispute the conclusions reached by the Bank staff, which the experts deemed sound.

Juba River. That is the Treaty between the Empire of Ethiopia and the Republic of Kenya respecting the Boundary between the two Countries signed in Mombasa on June 9, 1970. The Treaty has no bearing on the Baardhere Dam. Article VII of the Treaty provides that with respect to the Dawa River between Malka Rie and Malka Marie canalization of flood waters shall be permitted for cultivation, but no work shall be undertaken which might be prejudicial to the downstream water supply or which might alter the course of the river." Furthermore, the experts stated that both Ethiopia and Somalia are bound by the 1968 African Convention for the Protection of Nature and Natural Resources (*supra* n. 361) which requires prior consultation. For discussion of some of the legal aspects of the Report, *see* Caponera, *supra* n. 508.

[516] The independent experts noted further that "Regardless of the true figure, it appears unlikely that development could be at a rate which would create a negative factor during the economic life of the project. It could however, become a factor during the physical life of the project." *See* Report of Independent Experts, *supra* n. 514, at 2.

[517] *See id.*, at 3.

[518] *See id.*, at 8.

[519] The elements of a framework agreement were spelled out by the independent experts. They included establishment of an institutional mechanism, determination of priority actions on the river, a reciprocal acknowledgment of an equitable and reasonable utilization share in the use of the Juba River, and more importantly, determination of a time limit after which the Baardhere Dam would be constructed.

Accordingly, the independent experts confirmed the determination of the Bank that the project would not cause appreciable harm to the other riparians, and would not be harmed by other riparians' possible water use. However, the experts qualified this conclusion by citing two factors, the data made available to them, and lack of direct knowledge of the actual situation in Ethiopia. Although the possibility of a field visit was included in the terms of reference, no such a visit took place.

In connection with the recommendation for a framework agreement, the independent experts proposed that they be permitted to visit the appropriate ministry heads in Ethiopia, Kenya and Somalia to provide a vehicle for facilitation of negotiations between those riparians. This was clearly beyond their terms of reference and the role set forth for them under OMS 2.32, and was thus not accepted by the Bank.

The independent experts were no doubt eminent authorities in their fields. It is worth noting that they represented both, engineering and water law, the most relevant disciplines to the project and the objection. Clearly, the experts conducted their assignment in an independent environment, being selected by, and reporting to a unit other than the unit proposing the project.[520]

A few months after the report of the independent experts was submitted, the political and security situation in Somalia started to deteriorate, and as a result the Bank suspended processing of the project. Eventually, the situation in Somalia put an end to the processing of the Baardhere Dam Project. It would have been interesting to see how Ethiopia would have reacted to the decision of the Bank to proceed with the project, and how the Bank would have addressed any further protests from Ethiopia.

The Baardhere Dam Project in Somalia was the only project that has prompted the Bank to appoint independent experts to review an objection from one of the riparians to a project on an international waterway proposed for Bank financing. The subsequent objections that the Bank received from other riparians to some of its projects on international waterways were all dealt with by Bank management. In all of them the Bank decided to proceed with the project despite the objections because it did not find any merit warranting the appointment of independent experts. However, despite the incomplete story, the Bank and the independent experts in the Baardhere Dam Project have put in place an excellent road map for the selection and work of independent experts, if any were to be appointed again.

6.4   Bank Procedures for Dealing with Objections and Riparians' own Arrangements

The uses and sharing of a number of international watercourses are governed by agreements concluded by some of the riparians, and in few instances, such

---

[520] *See supra* n. 512.

agreements encompass all of the riparians to that watercourse. Most of these agreements include provisions on dispute settlement which vary in their scope, procedures and choice of forum.

Both the Treaty Concerning the Rio de la Plata 1973,[521] and the Statute of the River Uruguay 1975,[522] state that any dispute concerning the interpretation or application of either Treaty, which cannot be settled by direct negotiations, may be submitted by either of the parties to the to the International Court of Justice.[523] Similarly, the 1994 Danube Convention provides for the submission of any dispute arising between the parties to the International Court of Justice or through arbitration, if the parties fail to resolve the dispute through negotiations with the assistance of the ICPDR.[524]

The Mahakali Treaty concluded between India and Nepal in 1996[525] refers any difference that may arise between the parties, as a first step, to the Mahakali River Commission. If either party disagrees with the recommendation of the Commission, then the difference becomes a dispute, and such dispute would be submitted for arbitration to a tribunal consisting of three arbitrators. Each party would appoint one arbitrator, and the third arbitrator, who would preside over the tribunal, would be appointed jointly. If the two parties fail to agree on the

---

[521] *See supra* n. 362.

[522] *See supra* n. 363.

[523] *See* paragraph 87 of the 1973 Treaty and paragraph 60 of the 1975 Treaty. In this connection, on May 4, 2006, relying on the latter paragraph, Argentina brought a case against Uruguay before the ICJ alleging that Uruguay was in breach of the 1975 Treaty on the River Uruguay. The River Uruguay constitutes the boundary between the two countries. As a basis for the ICJ's jurisdiction, Argentina cited Article 60 of the 1975 Treaty, the first paragraph of which allows submission of a dispute to the ICJ if direct negotiations between the parties have not settled the dispute. The case relates to the construction by Uruguay of pulp mills across the River Uruguay which Argentina claims would damage the environment of the river. The case is still pending before the ICJ. *See Pulp Mills on the River Uruguay* (Argentina v. Uruguay), available at: http://www.icj-cij.org/docket/index.php?p1=3&p2=1&code=au&case=135&k=88. It should be added that although the river constitutes the boundary between the two countries, the dispute does not relate to the issue of delimiting the boundaries between them, *see supra* n. 51.

[524] Article 24 of the Convention lays down the situations for submission of a dispute to the ICJ. Annex V of Convention includes detailed provisions on arbitration.

[525] *See* Treaty Between His Majesty's Government of Nepal and the Government of India Concerning the Integrated Development of the Mahakali River Including Sarada Barrage, Tanakpur Barrage and Pancheshwar Project. For the full text of the Mahakali Treaty *see,* 36 I.L.M. 531 (1997). For a general discussion of the Treaty, *see* Salman M. A. Salman & Kishor Uprety, *Hydro-Politics in South Asia: A Comparative Analysis of the Mahakali and the Ganges Treaties,* 39 Natural Resources Journal, 295–343 (1999). *See* also Salman & Uprety, *supra* n. 61.

third arbitrator, either party may request the Secretary General of the Permanent Court of Arbitration to appoint such arbitrator.[526]

The 1992 Agreement between Kazakhstan, the Kyrgyz Republic, Tajikistan, Turkmenistan and Uzbekistan on co-operation in interstate water resources states that all disputes "have to be solved by the republican water-economic organizations heads and if necessary with participation of independent side's representative."[527]

As discussed earlier, the Indus Waters Treaty establishes a unique system for resolving issues that may arise between the two parties. Questions regarding interpretation and application of the Treaty are referred to, and decided by the Permanent Indus Commission. Differences are dealt with by a neutral expert, and disputes by a court of arbitration.[528]

The preceding Parts of this Chapter described in detail the procedures which the Bank would follow if one of the riparians objects to a project proposed for Bank financing. These procedures include, as discussed earlier, the possibility of appointing independent experts. A question that could arise in this connection is: how would the Bank handle an objection from one of the riparians to a project on an international waterway that is governed by an agreement which includes its own provisions for dispute settlement? Will the Bank follow its own procedures or will it adhere to the procedures set forth in the agreement between the parties, such as those outlined above?

The situation may become more complicated if the objecting state invokes the provisions of a treaty and refers the dispute to the entity specified in the agreement for resolution, while the Bank is still reviewing and deciding how to handle the objection. Should the Bank suspend the process and wait for the decision of this entity, or can it proceed with its own procedures for handling the objection?[529]

---

[526] *See* Articles 9 and 11 of the Mahakali Treaty. It is worth noting that the Mahakali Treaty uses the terms "difference" and "dispute" in the same manner used in the Indus Waters Treaty. *See supra* n. 227, and the discussion in Chapter 4, Part 4.2 of this Book.

[527] *See* Article 13 of the 1992 Agreement between the Republic of Kazakhstan, the Kyrgyz Republic, the Republic of Tajikistan, Turkmenistan and the Republic of Uzbekistan on co-operation in interstate water, *supra* n. 428.

[528] *See supra* n. 227, and the discussion in Chapter 4, Part 4.2 of this Book. As explained earlier, this is not a hierarchical structure for dispute resolution; rather, it is a jurisdictional one.

[529] It is worth noting that the Agreement between the United Nations and the International Bank for Reconstruction and Development concluded in 1947 authorizes the Bank to request advisory opinions from the ICJ. Article VIII of the Agreement states "The General Assembly of the United Nations hereby authorizes the Bank to request advisory opinions of the International Court of Justice on any legal questions arising within the scope of the Bank's activities other than questions relating to the relationship between

The Bank has not yet faced any such situation. Objections to its financed projects have been few, and none of them confronted the Bank with the question of choice between its own procedures, and the procedures set forth in a treaty to which the borrower and the objecting riparian are parties. It should be pointed out in this connection that the Bank policy for projects on international waterways requires the Bank to ascertain whether the riparians have entered into agreements or arrangements, or have established any institutional framework for the international waterway concerned.[530] It also requires confirmation in the PAD if the issues involved are covered by an appropriate agreement or arrangement between the beneficiary state and the other riparians.[531] As discussed earlier, the Bank would not finance a project that contravenes any international environmental agreement to which the member country concerned is a party. Accordingly, the Bank will have to review any such objection in the context of the agreement between the parties before it invokes the provisions of its policy. As stated before, the objecting riparians do not have veto power over a Bank-financed project. The Bank will proceed with financing a project if it determines through its internal process, with or without an opinion from independent experts, that the project will not cause appreciable harm to other riparians. However, if an agreement were to include provisions requiring the consent of other riparians to any project on the shared waterway, then the Bank will recognize such a requirement and proceed accordingly.[532]

As has been discerned throughout the Chapters of this Book, agreements on shared watercourses are increasing, and the dispute settlement provisions in these agreements are becoming more novel and complex. It is worth emphasizing in this regard that the Bank policy for projects on international waterways does not over-ride the riparians' obligations under any treaty or convention to which they are parties. Accordingly, the provisions on dispute settlement in any of those agreements would need to be carefully studied before a decision is taken on whether to follow the Bank's procedures on objection, or the procedures for dispute settlement set forth in that agreement.

---

the Bank and the United Nations or any specialized agency. Whenever the Bank shall request the Court for an advisory opinion, the Bank will inform the Economic and Social Council of the request." *See supra* n. 5.

[530] *See* paragraph 5 of OP 7.50.

[531] *See id.*, paragraph 8.

[532] Such a requirement would certainly have to be explicit and clearly spelled out, and the agreement would have to be recognized and accepted by all the parties.

CHAPTER 7

Exceptions to Notification under the Bank Policy

7.1 Early Practice

The initial flexible approach adopted by the Bank in the early 1950s for projects on international waterways required proposing and obtaining management approval of procedures for dealing with the international aspects of the project. As discussed earlier, each project on international waterways was dealt with according to its own facts and circumstances. That *ad hoc* approach continued until 1965 when OM 5.05 was issued adding some rules and procedures regarding presentation of the loans and credits to the Executive Directors. It may be recalled that OM 5.05 itself did not include explicit provisions on notification. However, the OM required informing the Executive Directors, *inter alia*, if the other riparians gave their no-objection to the project.[533] As discussed before, for the other riparians to give a no-objection, they need to be notified of the project that the Bank intended to finance. Thus, notification gradually became the main element of the Bank policy.

OM 5.05 stated that the information to be submitted to the Executive Directors should also indicate, *inter alia*, that "the project is not harmful to the interests of other riparians and their absence of consent is immaterial or their objections are not justified."[534] As discussed earlier, the phrase that "the project was not harmful to the interests of other riparians" was sometimes interpreted in isolation of the total content of the sub-paragraph to create an exception to the notification requirement, when the staff deemed the project not harmful to the interests of the other riparians. This issue was not clarified until OMS 2.32 was issued in 1985. That OMS included two explicit new exceptions to the notification requirement, clearly and explicitly indicating that those are the only exceptions allowed under the policy.

---

[533] As stated earlier, paragraph 3 of OM 5.05 of 1965 states that the SAR and the Report and Recommendation of the President should "state that the Bank/IDA has considered the international aspects of the project and is satisfied that: (i) the issues involved are covered by appropriate arrangements between the borrower and other riparians; (ii) the other riparians have stated (to the borrower or to the Bank/IDA) that they have no objection to the project; or (iii) the project is not harmful to the interests of other riparians and their absence of express consent is immaterial or their objections are not justified."

[534] *See id.*

However, the issuance of OMS 2.32 with the two exceptions to notification did not put to rest the issue of other exceptions to the notification requirement. As discussed in the previous Chapters, there were some suggestions within the Bank that in projects on international waterways where there was "no issue, no conceivable harm," notification was not required by the Bank policy. However, as explained earlier, this view is not supported by the provisions of any of the policies issued in 1985, or the revision added thereafter.[535] The fact that there were two exceptions to the notification requirement specified under the 1985 OMS, and that a third specific exception was added 10 years later, negates any argument for the legal validity of this general approach to exceptions. The discussion below elaborates on those exceptions to the notification requirement under the Bank policy.

7.2   Exceptions to the Notification Requirement

The 1985 OMS brought about the needed clarifications and elaborations to the Bank policy in a number of areas. It specified the waterways and projects to which it applied, and set forth the general rule for notification of all riparians and how, when and by whom it can be undertaken. It also laid down detailed procedures regarding the likely responses, and how to deal with each such response. As noted above, one element introduced by the 1985 OMS has been the inclusion of two exceptions to the notification requirement. These exceptions have been incorporated with minor changes in the various directives for Projects on International Waterways issued since that time. A third exception, to be also discussed in Part 7.2.3 of this Chapter, was added in 1994.

7.2.1   Rehabilitation of Existing Schemes

The first exception to the notification requirement for projects on international waterways relates to the rehabilitation of existing schemes under certain specified conditions.[536] The scheme has to be an on-going or existing one. The project would involve rehabilitation, construction or other changes which (a) would not adversely change the quality or quantity of water flows to the other riparians, and (b) would not be adversely affected by the other riparians' possible water use. The policy further clarifies that the exception applies only to minor additions or alterations to the on-going scheme. It does not cover works or activities that would exceed the original scheme, change its nature, or so alter or expand its scope and extent as to make it appear a new or different scheme.

---

[535] *See supra* n. 372. *See infra*, Part 7.2.3 of this Chapter for a discussion of the third exception to the notification requirement.

[536] *See* paragraph 7(a) of OP 7.50.

Paragraph 7(a) goes on to add the possibility of informing the Executive Directors concerned in case of doubt regarding the extent to which a project meets these criteria of the exception, and giving them at least two months to reply. Moreover, the paragraph states that even if the project meets the criteria for this exception, the Bank tries to secure compliance with the requirements of any agreement or arrangement between the riparians.

This is indeed a unique exception to the notification requirement. It does not have a corresponding equivalent in any of the other international instruments dealing with shared watercourses. As may be recalled, the IIL, the ILA and ILC have all set a threshold for notification of "serious effect," "material effect," or "significant adverse effects," respectively. Bank records do not shed any light on the reasons behind this exception. It seems that the decision to include this exception was influenced by the Bank operational work at that time. The Bank has been financing a large number of projects dealing mainly with rehabilitating and modernizing existing irrigation schemes, as well as water supply infrastructure. In most of those projects, it was argued that the water scheme or infrastructure would be operating more efficiently after the rehabilitation, and that could even result in water savings. The overflow of water from dilapidated canal banks and the unaccounted-for water that is lost through leaking pipes would be put to a good use when the system is rehabilitated and modernized. Even when minor additions were made to the existing scheme, as the exception allows, the system, in most cases, would, by and large, still be using no more water than it was using before rehabilitation. Apparently this line of argument prompted exemption of those projects from the general requirement of notification under the Bank policy.[537]

One important feature of this exception is that it allows a subjective judgment by the Bank staff that projects involving rehabilitation works which would not adversely change the quantity or quality of water flows to other riparians are exempt from the notification requirement under the policy. This judgment is not allowed in projects that involve new construction and not just rehabilitation. In projects involving new construction, notification would be required even if the project would not adversely change the quality or quantity of water flows to other riparians. Examples of such projects include small water supply components in a rural development or a community driven development (CDD) project, or minor flood works. In those types of projects there is a requirement under the policy of notification of all other riparians. The distinguishing factor is whether the works are for a new project, or for rehabilitation of an existing project. The works may

---

[537] Projects for the rehabilitation of existing schemes that would not cause adverse effects to other riparians would not be covered by the threshold for notification under the ILA and IIL rules, or the UN Watercourses Convention, as discussed above.

be more extensive in case of rehabilitation than in a new project. The cost may be the same, or it could even be higher in the case of rehabilitation. The Bank loan or credit may also be larger for a rehabilitation project than for a new project. Yet, under the policy notification would be required for the new works, but not for the rehabilitation works.

Typical projects that fall under this exception include the rehabilitation of the water infrastructure in irrigation or water supply projects. In one project in Kazakhstan, the works consisted of rehabilitation and modernization of the irrigation and drainage infrastructure, including main, inter-farm, and on-farm systems.[538] The project aimed to rehabilitate and bring the existing irrigation and drainage infrastructure back to fully operational condition, and allow for improved management of water and reduction in water losses. No new canals or structures that would result in an increase in the abstraction and supply of water were planned, and no development of new irrigation areas would be financed under the project. The project would thus not involve works or activities that would exceed the original scheme, change its nature, or alter or expand its scope and extent to make it appear a new or different system. The works under the original project abstract water from a number of rivers including the Ili River which Kazakhstan shares with China, the Chui and Talas Rivers which are shared by the Kyrgyz Republic and Kazakhstan, and the Syr Darya and some of its major tributaries which Kazakhstan shares with the Kyrgyz Republic, Tajikistan, and Uzbekistan, and which flows into the Aral Sea.

As discussed earlier, in 1992 the central Asian Republics of Kazakhstan, the Kyrgyz Republic, Tajikistan, Turkmenistan, and Uzbekistan have entered into an agreement on cooperation in the field of joint water resources management and conservation of interstate water sources.[539] However, it was concluded by the Bank that this agreement does not require notification of other riparian states for rehabilitation of existing schemes.[540] Accordingly, the rehabilitation project in Kazakhstan met the elements of the criteria for the rehabilitation of existing schemes under paragraph 7(a) of the Bank policy, and the existing agreements to which Kazakhstan is a party do not require notification of other riparians for the type of activities financed under this project.

Similarly, in a project in Azerbaijan, the Bank financed the rehabilitation of water supply and sewerage systems in some selected rayons, as well as of

---

[538] *See* Kazakhstan, Second Irrigation and Drainage Improvement Project (P086592, 2009).

[539] *See supra* n. 428.

[540] It was noted that Article 5 of the 1992 Agreement requires exchange of data and information on complex use and protection of water resources. However, it was concluded that this requirement does not apply to the use of water under the project.

water, wastewater and septic sludge treatment facilities, with the aim of providing better access to water supply and wastewater services, and improving water quality.[541] The water infrastructure system to be rehabilitated abstracts water from international rivers which include the Kura River which Azerbaijan shares with Georgia, and the Aras River which is shared by Azerbaijan, Iran and Turkey. The PAD for the project states that, not only would there be no adverse effects; rather the water intake from the rivers would actually be reduced because of the reduction in leakages and improved demand management under the project. The investments were also expected to improve the quality of wastewater discharged into the relevant waterways, resulting in overall improvements in their water quality. Because the activities to be financed under the proposed project consisted mainly of rehabilitation of existing schemes, it was concluded that the project would fall under the exception to the notification requirement under paragraph 7(a) of the policy. As mentioned before, Iran and the Soviet Union concluded an agreement in 1957 for the joint utilization of the frontier parts of the Aras and Atrak Rivers for irrigation and generation of power.[542] It was also concluded that the 1957 agreement does not require notification or exchange of information for the type of activities funded under the project.

The Infrastructure Project for the Province of Buenos Aires[543] included financing (a) for improvement and expansion of the secondary sewerage and drainage network which drain into existing primary networks and, after treatment, discharge into the Rio de la Plata; and (b) feasibility studies for similar future secondary sewerage and drainage investments, which may involve international waterways. The Rio de la Plata is an international waterway which Argentina shares with Uruguay. It was determined that the activities to be financed would deal with minor additions to the existing schemes, and would not cover works that exceed the original scheme, change its nature, or so alter or expand its scope and extent to make it appear a new or different scheme. It was further agreed that the activities would not adversely affect the quality or quantity of water flows to the other riparians; and would not be adversely affected by other riparians' water use. With regard to the feasibility studies which may involve international waterways, it was agreed that the terms of reference would include an examination of any potential riparian issues as required in paragraph 7(b) of OP 7.50 (discussed in the next Part of this Chapter).

---

[541] *See* Azerbaijan, Water Supply Project (P096213, 2007).

[542] *See supra* n. 146.

[543] Argentina, Infrastructure Project for the Province of Buenos Aires (APL2) (P105288, 2007).

The 1973 Treaty Concerning River Plata and the Corresponding Maritime Boundary concluded by Argentina and Uruguay[544] requires notification to the Administrative Commission only when the proposed activities are expected to cause adverse impacts on the navigability and/or the hydrological conditions of the Rio de la Plata. Accordingly, it was concluded that the nature of the works under the proposed project would not require notification under the Bank policy or under the 1973 Treaty.

The Electric Power Emergency Reconstruction Project in Serbia[545] involved improvements to a thermal power plant, including repairs of the boiler's super heater, turbine overhaul and small investments in the unit control system, and refurbishment of the plant's auxiliary systems. The project is located on the right bank of the Danube River, about 20 kilometers from the Romanian border. In normal operation, the plant uses 7.6 cms (cubic meters per second) of cooling water from a channel connected to the Danube; all such water is returned to the Danube at a slightly higher temperature. A question arose as to whether the rise in temperature of the returned water could have adverse effects on the Danube River, or any of its riparians. It was noted that the rise of temperature at the channel outlet to the Danube is practically 0°C due to the small quantity of cooling water compared to the overall Danube flow (which averaged 5,600 cms). Accordingly, it was concluded that the activities under the project were of rehabilitation nature and would not have adverse effects on the quality or quantity of water flow to other riparians, and thus would fall under the exception to the notification requirement of paragraph 7(a) of the Bank policy. It was further noted that the existing agreements to which Serbia is a party do not require notification for such kinds of activities.[546]

Another feature of this exception worth noting is the requirement that the project will not adversely change the quality or quantity of water flows to other riparians. The clause "adversely change" contrasts with the clause "cause appreciable harm" that is used a number of times in the policy in connection with new construction. In fact, the clause "adverse change" is only used in relation to this exception. This difference in terminology used raises the question of whether

---

[544] *See supra* n. 362.

[545] Serbia, Electric Power Emergency Reconstruction Project (P074136, 2001).

[546] Two treaties were identified and reviewed: (i) Convention Concerning the Regime of Navigation on the Danube (with annexes and supplementary protocol), between Union of Soviet Socialist Republics, Bulgaria, Czechoslovakia, Hungary, Romania, Ukraine Soviet Socialist Republic and Yugoslavia, signed at Belgrade on August 18, 1948; *see* 518 U.N.T.S. 197 (1949); and (ii) Protocol (with Annexes) Governing Crossing of the Frontier by Officials of the Water Control Services, between Yugoslavia and Romania, concluded on December 31, 1948.

under the policy "adverse change" is understood to be less significant than "appreciable harm." That may be the reason, although there is nothing in the Memorandum on Riparian Rights,[547] or in the policy explaining the difference.[548] There is also the possibility that those two terms are simply being used interchangeably. However, the requirement under this exception is that the Bank would not finance a project involving the rehabilitation of existing schemes that would cause adverse effects to other riparians.

A question has arisen in connection with projects that involve both construction of new activities, and rehabilitation of existing schemes. Should the notification letter and Project Details include a reference to the rehabilitation components, or be limited to the new activities under the project? Although the rehabilitation components fall under the exception to the notification requirement, the notification letters and Project Details have consistently included a detailed description of all the components of the project, including the rehabilitation ones. This is done to ensure full disclosure and transparency, and to assure the notified states that the cumulative effects of the new components and the rehabilitated components will still not cause any adverse effects to the other riparian states.

A third feature of this exception is that, in case of doubt regarding the extent to which the project meets the criteria of this exception, the policy requires that the Executive Directors representing the riparians concerned are informed and given at least two months to reply. As indicated earlier, the Bank did undertake notification in its early years in some instances through the Executive Director concerned.[549] The 1985 OMS has limited that approach only to this exception.

The Tarim Basin Project in China involved the financing of sub-projects that consisted of rehabilitation of main and secondary systems through the lining of canals, and the addition of minor tertiary irrigation and drainage systems to develop some additional dry lands.[550] The sub-projects were located along the Aksu River which originates in the Kyrgyz Republic, and the Kashgar River which originates in the Kyrgyz Republic and Tajikistan, before both flow into China. Although there was agreement that the bulk of the project activities fell under the exception to the notification requirement of paragraph 7(a), rehabilitation of existing schemes, the addition of the minor tertiary irrigation generated a debate as to whether the project would still be covered by this exception. There

---

[547] *See supra* n. 151.

[548] For the ILC definition of the term "appreciable" *see supra* n. 345.

[549] Direct notification by the Bank to the riparian states would not prevent the notified state from responding to the notification through its Executive Director, as happened in a number of cases, including the Igdir-Aksu Project, *supra* n. 143.

[550] *See* China, Tarim Basin II Project (P046563, 1998).

was also discussion as to whether those activities would meet fully the parameters of the exception, and that the Executive Directors representing Kyrgyz Republic and Tajikistan might need to be informed of the project as per paragraph 7(a) of the policy. However, it was agreed that the additions were minor and would not exceed the original scheme or change its nature, and that the additional water needs for the new components would come from the water savings generated as a result of the rehabilitation of the existing scheme. It was accordingly concluded that there would be no adverse effects on any of the riparians of either river, and that the project would fall under the exception to the notification requirement under paragraph 7(a) of the policy. Hence, the requirement under the paragraph informing the Executive Director was not invoked.

One of the questions raised in connection with this exception relates to rehabilitation of existing dams. Given the controversies that surround most of the dams built on international rivers, should the rehabilitation of existing dams still be covered under this exception? Bank-financed projects in this area concentrated on the rehabilitation and improvement of spillways, head regulators, draw-off gates and their operating mechanisms. They also addressed leakages and seepage, and included improvements in the ability of the dam to withstand higher floods. None of the projects processed under this exception included increases in the height of the dams or in the reservoir capacity.[551] This approach to rehabilitation of existing dams under the exception of paragraph 7(a) seems to address both the letter and spirit of the exception, and has raised no concerns from the other riparians.

Paragraph 7(a) of the policy has used the term "informed" and not "notified" in connection with the conveyance of information about the project and the exception to the Executive Directors concerned. It does not seem that from a practical point of view that there would be any difference in the results if one or the other term is used. The Executive Directors would be given at least two months to reply. However, the Executive Directors would most likely not take the decision themselves. Rather, the matter would in most cases be referred to the governments of the riparians concerned to study themselves, and make their own determination. This would open the door for any of the possible responses discussed in the previous Chapter of this Book (such as consenting to the project,

---

[551] *See* as an example, India, Dam Rehabilitation and Improvement Project (P089985, 2009). The Project support included treatment of leakage and reduction of seepage, improving dam drainage and ability to withstand higher floods, including additional flood handling facilities, rehabilitation and improvement of spillways, head regulators, draw-off gates and their operating mechanisms. The support also included hydrological assessments, sediment management, and other measures required to improve the safety and operation of the dams and associated appurtenances. However, the project did not include any increase in the dam height or the reservoir capacity.

requesting more information, more time, or both, or even objecting).[552] However, none of these eventualities has happened thus far, because no such doubtful cases, regarding the rehabilitation of existing schemes, have arisen.

Even if the project meets the criteria of this exception, paragraph 7(a) of the policy requires that the Bank try to secure compliance with the requirements of any agreement or arrangement between the riparians. This means that if the riparians have entered into an agreement that requires notification or exchange of data and information for those types of projects, then the Bank should try to ensure that the provisions of such agreement are complied with. As discussed above, there are agreements in place for some of the international waterways affected by each of the projects in Kazakhstan and Azerbaijan. Those agreements do not require notification or exchange of information for the type of activities funded under each of those projects.

However, there are agreements that oblige the parties to exchange data and information on any activities on the shared river in question, including rehabilitation works that are exempt from notification under the Bank policy. The Bank has taken note of such obligation, and has asked that the borrowers should comply with that requirement, notwithstanding the fact that the project falls under the exception to the notification requirement under the Bank policy. This is because the Bank policy does not supersede the borrower's obligation under an international treaty.

Indeed, OMS 2.20 on Project Supervision issued in January 1984 stated that: "Should international agreements exist that are applicable to the project and area, such as those involving the use of international waters, the Bank should be satisfied that the project loan is consistent with the terms of the agreement."[553] A few months later, in May 1984, the Bank issued its first OMS 2.36 on the environment.[554] That OMS went one step further and stated that: "the Bank will not finance projects that contravene any international environmental agreement to which the member country concerned is a party."[555] It may be argued that OMS 2.36 limits the Bank obligations under OMS 2.20 only to international environmental agreements. However, it is clear that the obligation under OMS 2.36 is more categorical than that under OMS 2.20. The obligation of the Bank not to finance projects that contravene the borrower's international environmental agreements

---

[552] *See supra* Chapter 5, Part 5.5.

[553] *See* paragraph 24 of OMS 2.20, "Project Supervision." It is worth noting that OMS 2.20 was issued in January 1984, almost a year and a half before OMS 2.32, Projects on International Waterways, was issued, in April 1985.

[554] *See* OMS 2.36, Environmental Aspects of Bank Work, *supra* n. 109.

[555] *See* paragraph 9(c) of OMS 2.36.

has been included in all the Bank directives on the environment issued since 1984, and is also included in the current Bank policy on "Environmental Assessment."[556]

One example of treaties that set forth a more stringent requirement for notification and exchange of data and information is the 1980 Convention Creating the Niger Basin Authority (NBA).[557] The Convention requires the riparian states to keep the Executive Secretariat of the NBA informed of all projects and works they intend to carry out in the Basin.[558] This obligation has necessitated notifying or informing the Executive Secretariat of the NBA of all Bank-financed projects on the Niger River. In projects involving new construction, notification has been undertaken for some projects by the borrowers and in others by the Bank, on behalf of the borrowers.[559] However, when the project involves only rehabilitation of existing schemes that meet the criteria of the exception under the policy, the Bank has taken the position that there is a requirement under the NBA Convention to inform the NBA Executive Secretariat, and that informing the Secretariat should be done by the borrower, in fulfillment of its obligations under the 1980 Convention.[560]

---

[556] OP 4.01, Environmental Assessment (EA) (issued in January 1999), states in paragraph 3 that the "EA considers natural and social aspects in an integrated way. It also takes into account the variations in project and country conditions; the findings of country environmental studies; national environmental action plans; the country's overall policy framework, national legislation, and institutional capabilities related to the environment and social aspects; and obligations of the country, pertaining to project activities, under relevant international environmental treaties and agreements. The Bank does not finance project activities that would contravene such country obligations, as identified during the EA." This same obligation is included in OP 4.36, Forests (2002). Paragraph 6 of this OP states that "the Bank does not finance projects that contravene applicable international environmental agreements."

[557] *See* Convention Creating the Niger Basin Authority, *supra* n. 247.

[558] Article 4(4) of the Convention states that: "The Member States pledge to keep the Executive Secretariat informed of all the projects and works that they might intend to carry out in the Basin. Moreover, they pledge not to undertake any work on the portion of the River, its tributaries and sub-tributaries under their territorial jurisdiction which pollute the waters or modify the biological features of the fauna and the flora."

[559] The Nigeria, Local Empowerment and Environmental Management Project (P069892, 2003) included the construction of small water harvesting structures, small scale irrigation schemes, diversion ditches, protection bunds, drinking water wells, boreholes, and the protection of natural springs. Those micro-projects would be developed, *inter alia*, on tributaries or sub-tributaries of the Niger River. Nigeria notified the NBA in October 2000, and asked the NBA to inform the member states of the NBA of the project.

[560] The Nigeria, National Urban Water Sector Reform Project (P071075, 2004) involved the rehabilitation of the existing water distribution network, drawing water from the Niger River in five states in Nigeria. The Bank processed the project under the exception to the notification requirement and missed invoking the requirement of the 1980 NBA Convention. This omission was brought to the attention of the Bank by an NGO working on the Niger Basin. As a result the Bank asked Nigeria to inform the NBA of the project, pursuant to the 1980 NBA Convention. This was eventually done. The fact that an NGO

Another example is the Agreement between the Government of the Republic of Croatia and the Government of Bosnia and Herzegovina on Regulating Water Management Relations concluded on December 26, 1966. The Agreement calls for close collaboration on all areas of water management, and requires the two parties to resolve by consent all water management issues particularly those relating to water use, protection, and maintenance of the water facilities.[561] Based on those provisions of the Agreement, the Bank decided that it would be necessary for Bosnia to notify Croatia of the Mostar Water and Sewerage Project despite the rehabilitation nature of the works financed under the project. Indeed, the project focus was rehabilitation of the water and sanitation system in the city of Mostar.[562] The City is located on the banks of the Neretva River which originates in Bosnia and Herzegovina, and flows for a short distance in Croatia before emptying into the Adriatic Sea. Bosnia notified Croatia, and the latter responded expressing support for the project.

Some agreements address more complex issues than a simple exchange of information, and accordingly would require extra attention from the Bank, even when the project involves the rehabilitation of existing schemes. One example is when the facilities to be rehabilitated are located outside the territories of the borrower.[563]

---

brought the matter to the attention of the Bank indicates the world-wide attention being paid to international waterways, and the high degree of knowledge of the Bank safeguard policies by a number of civil society organizations. *See* also Nigeria, Second National Urban Water Sector Reform Project (P071391, 2005).

[561] Article 2 of the Agreement states: "Taking into account the need of maintaining uniformity of water management relations, the Contracting Parties shall on the basis of the provisions of this Agreement resolve by consent all water management issues, particularly those ones regarding preparation of the project documentation and carrying out activities relating to water usage, preventing water pollution, protection from adverse water effects, maintaining water management facilities, reconstruction of the war damaged water management facilities and construction of new ones, and jointly regarding provision of financial resources from local and international sources..."

[562] Prior to the war in this region of former Yugoslavia, there was one water and wastewater utility serving Mostar, a city with a Bosnian and Croatian population. As a result of the war, the utility was split into two parts: the Eastern utility serving the Bosnian population, and the Western utility serving the Croatian population. The project aimed at the reunification of the two systems in order to achieve institutional, financial and economic benefits. *See* Bosnia and Herzegovina, Mostar Water Supply and Sewerage Project (P057951, 2000).

[563] The Uzbekistan, Karshi Pumping Cascade (KPC) Rehabilitation Phase I Project (P057903, 2002) aimed at the rehabilitation and modernization of the KPC, which draws water from the Amu Darya River. As discussed earlier, Uzbekistan shares this river with Tajikistan, Turkmenistan, and Afghanistan, and because the river flows into the Aral Sea, Kazakhstan is also a riparian. The project consisted of the rehabilitation of existing schemes, which would not cause adverse effects to the quantity or quality of water flows to other riparians, and hence would fall under the exception to the notification requirement under paragraph 7(a) of OP 7.50. However, most of the facilities to be rehabilitated

These are just a few examples of agreements that are more stringent in the requirement of exchange of data and information than the Bank policy. Still, the Bank has paid particular attention to this matter, and ensured that the requirements under these agreements are met. In addition to the fact that the Bank policy does not supersede the borrower's obligations, one of the Bank's overarching objectives has been dispute avoidance. This objective would only be achieved through ensuring that the riparians' rights and obligations are fully taken into consideration in projects that may affect them.

In this connection, one question that has arisen is whether the Bank would finance the rehabilitation of existing schemes that are clouded by controversy between the borrower and some other riparians, and which were originally not financed by the Bank. One argument for supporting the financing of these kinds of projects is that those projects deal with the rehabilitation of existing schemes, and as such are exempt from the requirement to notify other riparians.[564] However, the requirement for the Bank to act prudently in the interests of all the riparians, and the need for dispute avoidance are weighty arguments against financing such projects. As a result, the Bank has extended its rule against financing new projects on international waterways where a dispute exists to projects involving the rehabilitation of the existing schemes as well.[565]

---

are actually located and established in Turkmenistan through two agreements between Turkmenistan and Uzbekistan. Hence, Turkmenistan was notified of the project, and its permission sought and granted before the project commenced.

[564] This question arose in 1995 in connection with a request for advice whether the Bank can finance rehabilitation works on the Farakka Barrage. This barrage was constructed by India in the 1970s to divert some of the waters of the Ganges River to Calcutta city. Pakistan, and later Bangladesh, protested vehemently against the project as it would deprive Bangladesh of the waters of the Ganges River, and cause Bangladesh significant adverse effects. However, in 1996, one year after this issue of financing the rehabilitation works related to the Farakka Barrage arose, India and Bangladesh concluded a treaty for sharing the waters of the Ganges. For the history of the dispute over the Farakka Barrage, and discussion of the Treaty, *see* Salman & Uprety, *supra* n. 61, at 125.

[565] Some of the difficult questions that the Bank has faced go beyond whether it can finance rehabilitation works for existing controversial projects. One question relates to whether the Bank can finance works that are not directly connected with such a controversial project. For example, state A builds a dam on a river shared with state B, without notification of, or consultation with state B. State B alleges that the dam is causing significant adverse effects to it, and complains publicly about the dam. Can the Bank finance, at the request of state A, housing, roads, schools and hospitals in the area of the dam in state A? State B could ask the Bank not to finance such works because the financing would give legitimacy to the dam, and expand use of water from the shared river through the new infrastructure. The World Commission on Dams (WCD) that was established in 1997 to review the development effectiveness of large dams and assess alternatives for water resources and energy development has addressed this matter. The WCD issued its report in 2000; *see Dams and Development: A New Framework for*

The exception to the notification requirement under the policy dealing with the rehabilitation of existing schemes is no doubt a unique one. It gives the Bank the authority to make the determination, subject to the parameters specified in paragraph 7(a), that such works will not adversely change the quantity or quality of water flows to other riparians. Yet, notification is required in projects involving new construction even if such works would not adversely change the quantity or quality of water flows to other riparians. Similar to other provisions of the policy, and as the above discussion has shown, implementation experience has clarified a number of elements of this exception.

7.2.2 Water Resources Surveys and Feasibility Studies

The second exception to the notification requirement under the Bank policy covers water resources surveys and feasibility studies on or involving international waterways. No notification of other riparians is required for the carrying out of either of those surveys or studies under any Bank-funded project or technical assistance.[566] However, the policy requires that the state proposing such activities includes in the terms of reference for the study an examination of any potential riparian issues.

These kinds of studies are usually carried out as part of small technical assistance projects, although in a few instances they may be part of larger projects. Water resources surveys involve in most instances an assessment of the water flow and its different sources and uses, as well as other abstractions and storage losses, including evaporation. A feasibility study (sometimes referred to as pre-feasibility study) for a project typically includes a preliminary assessment of the environmental and social impacts and possible mitigation measures, dam safety issues, if any dams are involved, and potential riparian issues.

Decision-Making (The Report of the World Commission on Dams, Earthscan Publications, 2000). The Report addressed technical, economic, financial, and environmental aspects of dams, as well as the rights and obligations of riparian states over the shared rivers where the dams are built. The WCD's seven strategic priorities include "sharing rivers for peace, development and security," and its 26 guidelines include "procedures for shared rivers." Those procedures include five recommendations. The fifth recommendation of the WCD deals with dams on shared rivers that are built in contravention of good faith negotiations between riparians, including refusal to establish a review panel, or rejecting its findings when it is established. In such cases the WCD recommends that external financing agencies, whether bilateral, multilateral or export credit agencies, withdraw their support for projects and programs promoted by that agency. *See id.*, at 309. For more on the WCD, *see*: http://www.dams.org/. For the World Bank response to this and the other Guidelines of the WCD, *see*: http://go.worldbank.org/EGLG79W910.

[566] *See* paragraph 7(b) of OP 7.50.

One project that has dealt with this exception is the Afghanistan Water Sector Capacity Building Project.[567] The project aims to build the capacity of the Government of Afghanistan to progressively undertake strategic planning and to improve project preparation for water resources development. The Project includes water resource surveys, basin planning and a small number of pre-feasibility and feasibility studies for potential projects, mostly on the international rivers that Afghanistan shares with some of its neighbors, including the Kabul River and the Helmand River.[568] It was concluded that those activities fall under the exception to the notification requirement of paragraph 7(b) of the policy relating to water resources surveys and feasibility studies.

Water resources surveys and feasibility studies should be distinguished from the detailed design and engineering studies. As discussed earlier, the policy requires notification of the other riparian states when the Bank is financing detailed design and engineering studies for the projects specified under the policy. Although it would be possible as a general rule to distinguish between the two types of studies, there may be cases requiring extra caution.

This issue was discussed in the Greater Mekong Sub-region Power Trade Projects.[569] The Projects aim to enhance regional power trade within the Greater Mekong Sub-region, and include a component for the preparation of feasibility studies for hydroelectric projects involving the Mekong River and its tributaries. As discussed earlier, the Mekong River is shared by Cambodia, China, Lao PDR, Myanmar, Thailand and Vietnam. The studies include engineering, economic, and environmental and social assessments. It was determined that these components comprise feasibility studies and not detailed design and engineering studies, and thus fall under this exception to the notification requirement.

Conversely, the activities to be financed may relate only to the development of modeling tools to identify water-related investments and evaluate them in a regional context. It was concluded that this type of activity is not covered by the policy, even under the exception dealing with water resources surveys and feasibility studies.[570]

---

[567] *See* Afghanistan, Technical Assistance for Water Sector Capacity Building Project (P112097, 2009).

[568] The Kabul River originates in Afghanistan and flows into Pakistan where it joins the Indus River. *See* Salman & Uprety, *supra* n. 61, at 37. The Helmand River originates in Afghanistan and flows for a considerable distance before forming the borders for some stretches between Afghanistan and Iran, and then empties into the Helmand Swamps in Iran. Afghanistan and Iran have a long history of dispute over the River. A number of awards by third parties have been issued, starting with the 1872 Award. For more details *see* McCaffrey, *The Law of International Watercourses, supra* n. 185, at 236.

[569] *See* Greater Mekong Subregion Power Trade Phase 1 Project (Cambodia, P105329, 2007) and (Lao PDR, P105331, 2007).

[570] *See* Egypt, Ethiopia and Sudan, Eastern Nile Planning Model Project (P103639, 2009).

One observation about this exception is that it may be interpreted and applied as encouraging country specific water resources surveys, as opposed to basin-wide studies that would establish and strengthen cooperation among all riparians, and would deal with the shared waterway as one hydrologic unit. As a general rule, the Bank has followed a basin-wide approach to projects and programs which would include all the riparians, and thus would not require an exception to the notification requirement under this sub-paragraph. Such programs include "reversing land and water degradation trends in the Niger River Basin,"[571] and a similar project for Lake Chad Basin.[572] In addition to rivers and lakes, transboundary groundwater was also addressed in the Environmental Protection and Sustainable Development of the Guarani Aquifer System Project,[573] as well as the Groundwater and Drought Management Project in the Southern African Development Community (SADC).[574]

Notwithstanding the fact that feasibility studies fall under the exception to the notification requirement under the policy, the Bank ensures that the international aspects of projects on international waterways are dealt with at the earliest possible opportunity. Thus, when the feasibility studies concern large or complex projects, the Bank would ensure that other riparians are consulted and data and information are provided to them. The feasibility study for the proposed Rogun Hydroelectric Power Project in Tajikistan, with which the Bank is assisting, is an example. The proposed project involves construction of a dam, more than 300 meters in height, on the Vakhsh River with a reservoir of storage volume of about 13 cubic kilometers. The Vakhsh River is a tributary of the Amu Darya River which Tajikistan shares with Turkmenistan, Uzbekistan, and Afghanistan. Since the Amu Darya flows in the Aral Sea, Kazakhstan is also a riparian. As discussed earlier, the five riparian states concluded an agreement in 1992 on cooperation on interstates water resources.[575] Article 8 of the Agreement requires the parties not to allow any action which could result in a breach to the interests of the other riparians. Moreover, the Agreement establishes the Inter-state Commission for Water Coordination (ICWC) whose mandate includes elaboration and affirmation of water consumption limits annually for each party and the strict observance of release regime and water consumption

---

[571] *See* Reversing Land and Water Degradation Trends in the Niger River Basin (P070256, 2004), *supra* n. 247.

[572] *See* Reversal of Land and Water Degradation Trends in Lake Chad Basin Ecosystem (P070252, 2003), *supra* n. 246.

[573] *See* Environmental Protection and Sustainable Development of the Guarani Aquifer System Project (P068121, 2002), *supra* n. 241.

[574] *See* Groundwater and Drought Management Project (P070547, 2005), *supra* n. 248.

[575] *See supra* nn. 428 & 527.

limits. The ICWC mandate also includes appropriate scheduling of water reservoir operation. Under those circumstances, the terms of reference for the feasibility study of the proposed project have been shared with all the above riparians, and consultations have taken place with those countries on the proposed project.

Another feasibility study where the other riparians have been either briefed or consulted concerns the Red Sea–Dead Sea Water Conveyance Feasibility Study and Environmental and Social Assessment.[576] The falling level of the Dead Sea has become a major concern to the international community, and has been a topic of discussion in a number of international water conferences.[577] At the expressed jointly signed request from Jordan, the Palestinian Authority and Israel, the Bank is managing the financial and technical aspects of a feasibility study to determine what can be done to save the Dead Sea from further declines in its level. Conveyance of water from the Red Sea (at Aqaba) to the Dead Sea is the option proposed by the pre-feasibility study undertaken in the late 1990s, and is being promoted by the three parties at this time. The feasibility study would consider environmental, economic, technical, social, and financial aspects of the proposal. Although this is a feasibility study that falls under the exception to the notification requirement, the other two littoral states to the Gulf of Aqaba, Egypt and Saudi Arabia, are being consulted, and information about the study is being publicly disclosed and updated periodically.[578] This is being done because of the substantial scope and considerable visibility of the feasibility study.

Thus, similar to the first exception to the notification requirement dealing with the rehabilitation of existing schemes, the Bank is also ensuring that the will of the riparians, expressed in an agreement between them, is respected. The Bank has

---

[576] For the history and components of the Feasibility Study *see:*http://web.worldbank.org/WBSITE/EXTERNAL/COUNTRIES/MENAEXT/EXTREDSEADEADSEA/0,,content MDK:21827416~pagePK:64168427~piPK:64168435~theSitePK:5174617,00.html.

[577] The Jordan River is the main surface water body feeding the Dead Sea, but since most of the waters of the Jordan River are being used for water supply and irrigation activities, little or no water is reaching the Dead Sea. This situation has led to the significant decline of the Dead Sea and the environment of the Basin. The Dead Sea is the lowest spot on earth, about 392 meters below the sea level in the 1960s, and has fallen to about 418 meters below sea level in 2006; a drop of about one meter a year in the last 10 years. As a result of this drop, the Sea's water area has been reduced by one third—from about 950 square kilometers, to about 630 square kilometers. The Dead Sea is also the saltiest water body on earth—10 times more saline than ocean water. The Sea has created a wildlife and ecology based on the unique fresh-water, salt-water interface. For more on the problems of the Dead Sea and the proposed study *see id.* For an analysis of the political geography of the Jordan River, *see* Miriam R. Lowi, *Water and Power—The Politics of a Scarce Resource in the Jordan River Basin* (Cambridge University Press 1995).

[578] *See supra* n. 576.

also acted prudently and involved other riparians when it finances major feasibility studies, and has stayed away from controversial projects even when the activities are purely rehabilitation in nature.

7.2.3 Projects in a Tributary Exclusively in the Lowest Downstream Riparian

The third exception to the notification requirement deals with projects in a tributary that runs exclusively in the territory of the lowest downstream riparian.[579] Unlike the two exceptions discussed above, this exception was not included in the 1985 OMS. It was added to the policy in 1994, following discussion of a project in Burma (currently Myanmar).[580] The project consisted of the construction of an earth dam about 110 feet high on the Yin River, and diversion structures about 17 miles downstream of the dam. The Yin River originates within Burma and joins the Irrawaddy River there. The Irrawaddy River is the largest river in Burma. It is fed by two tributaries, the Daying and the Jiang, both originating within China, with the latter tributary fed by a sub-tributary originating within India. Burma is the lowest downstream riparian of the Irrawaddy River, and from Burma, the River empties into the Indian Ocean.[581]

The international waterway aspect of the project centered on the fact that the project is on a tributary originating within the lowest downstream riparian, and joining an international waterway, the Irrawaddy River, there. Although fed by some tributaries originating in China and India, the Irrawaddy River itself runs largely in Burma. It was argued strongly that the project would not cause appreciable harm to the other riparians, nor would it be appreciably harmed by the other riparians' possible use of water. This is because the Yin River runs only in Burma, the lowest downstream riparian. Based on this argument, a decision was taken that since there was no conceivable harm, there was no issue,[582] and accordingly notification of other riparians was not required.

Subsequently, a third exception was added and included in the 1994 version of the Bank policy as paragraph 7(c). The new exception to the notification requirement covers "any project that relates to a tributary of an international waterway where the tributary runs exclusively in one state and the state is the

---

[579] *See* paragraph 7(c) of OP 7.50.

[580] *See* Burma, Second Tank Irrigation Project (1988).

[581] The name of the country was changed from Burma to Myanmar in 1989. More than 90 percent of the catchment area of the Irrawaddy River is in Myanmar.

[582] Although the approach of "no conceivable harm, no issue" has no basis in OMS 2.32 of 1985, it prevailed in this particular case, as discussed by Goldberg, *supra* n. 372.

lowest downstream riparian, unless there is concern that the project could cause appreciable harm to other states."

This exception seems to have been influenced by the fact that the Irrawaddy River is largely a Burmese river, with only a small portion of the catchment area and the basin falling outside of Burma. However, the exception overlooked two facts. First, it failed to treat the Irrawaddy River and its tributaries as one basin where cooperation and goodwill of all the riparians is required for its efficient utilization. As discussed earlier, the definition of the term "international waterway" under the Bank policy includes tributaries of such a waterway.[583] Second, the exception failed to take into account the fact that the project would assist Burma in establishing rights to the waters of the Yin River obtained as a result of the project, and that such water rights would be claimed as established or acquired rights in any future negotiations with the other riparians. The waters which Burma would claim as established rights would not be available for the future uses of the other riparians of the Irrawaddy basin. This concept of foreclosure of future uses has been discussed earlier.[584] Another related point is the fact that the Irrawaddy River flows into the Bay of Bengal. No discussion took place as to whether the Bay of Bengal would qualify as a "bay" under the policy, requiring notification of the littoral states.[585]

It is worth noting that this exception to the notification requirement used the term "appreciable harm" in line with the rest of the policy, but unlike the first exception discussed above which used the term "adverse effect." In this way, the two exceptions under the same paragraph of the policy have used two different terminologies for the threshold for the activities financed to fall under said exception. Like the first exception, there is a qualification for the exception that the project would not cause appreciable harm to other riparians. It is worth noting that the paragraph refers to concern about harm, and not to actual harm.

Another observation about this exception is the use of the term "downstream." As may be recalled the 1989 OD used the terms "downstream and upstream" but that version of the policy was soon withdrawn and the new version of the policy reverted to the term "other riparians" in place of those terms. In this particular exception there was a clear need to use the term "downstream riparian" as the exception is referring to a specific riparian on a specific location on the shared river.

This exception has rarely been invoked in any operation since it was added to the policy in 1994.[586] In fact, the project which prompted the discussion on

---

[583] *See* paragraph 1(b) of OP 7.50.

[584] *See supra* Chapter 5, Part 5.3 of this Book.

[585] *See supra* n. 274.

[586] The exception was discussed in connection with some projects in Nigeria, on some of the tributaries of the Niger River that originate within Nigeria, the lowest downstream

this exception, and later resulted in adding it to the policy later, itself never materialized.[587]

Of the three exceptions, the first one relating to the rehabilitation of existing schemes is the most widely resorted to. As discussed earlier, some borrowers would not want to notify other riparians for projects on waterways shared with them, and the exception provides a way out of notification. Some Bank staff who are concerned about the requirements of notification, and are apprehensive about objections, prefer to concentrate on rehabilitation of existing schemes. In this way, however, the exception might have created a distorted incentive for staying away from new projects on international waterways.

The Bank has put in place detailed procedures regarding any of these exceptions. A memorandum needs to be prepared by the Task Team and sent to the Regional Vice President (RVP), describing the proposed works (or studies), the international waterways on which they fall, the countries sharing the waterway, and explaining how the works (or the studies) are covered by the exception. The memorandum would also confirm that the works will not adversely change the quality or quantity of water flows to the other riparians, and will not be adversely affected by the other riparians' possible use of water. This would be evidenced by the figures for any amounts of water that may be used or saved. Furthermore, the memorandum would refer to any agreements on the waterway in question to which the borrower is a party, and confirm that such agreements do not require notification or exchange of data or information on the types of works (or studies) to be financed under the proposed project. If an agreement exists that requires the exchange of data and information (as opposed to notification), the memorandum would explain how the borrower's obligations under this agreement have been met.[588] The RVP usually approves the Task Team's determination on the exception. This memorandum is ordinarily prepared and sent immediately after

---

riparian. However, since water is being abstracted also from the main Niger River, the notification letters for those projects included both the main river as well as those tributaries originating within Nigeria. Thus, no exception for notification under sub-paragraph 7(c) was sought for those tributaries. *See* Nigeria, Local Empowerment and Environmental Management Project (P069892, 2003).

[587] Processing of the project was halted after negotiations, following discontinuance of the Bank operations in Burma in 1988.

[588] Although the policy does not include explicit provisions requiring this memorandum, the practice of preparing this memorandum is based on the need for a record of how the decision has been made, and is inferred from the general requirements under paragraph 1 of BP 7.50, *supra* n. 462. As discussed earlier, the paragraph requires describing the international water rights issues in all project documents, starting with the Project Information Document (PID). It also requires the Task Team (TT) to prepare the project concept package to convey all relevant information on international aspects of the project, and states that throughout the project cycle the Region, in consultation with LEG, keeps the Managing Director (MD) concerned abreast of the international aspects of the project and related events.

appraisal when most of the information, including its description and environmental impacts, are well known. This course of action establishes a clear record of the process for the determination that the project falls under one of the exceptions to the notification requirement.

The above discussion has shown that of the three exceptions to the notification requirement under the policy, only the one relating to the rehabilitation of existing schemes is widely used. The second exception is rarely used, and the third one has actually not been used at all. The fact that the first exception allows Bank staff the space to determine that certain types of projects will not adversely change the quality or quantity of water flows to other riparians has prompted discussion about a possible wider exception. This wider exception would give the Bank staff the ability to determine if any project, and not just projects dealing with the rehabilitation of existing schemes, will not adversely change the quality or quantity of water flows to other riparians. If the project is not expected to do so, then the rationale for the exception would be extended to it. In other words, any project that does not adversely change the quality or quantity of water flows to other riparians would be exempt from the requirement of notification. This proposal will be addressed in Chapter 10 of this Book.

CHAPTER **8**

The Bank Policy and Transboundary Groundwater

8.1   The ILA Work on Transboundary Groundwater

The Helsinki Rules issued by the ILA in 1966, as discussed earlier, were the first international legal instrument to address transboundary groundwater. The Rules which deal with international drainage basins define such a basin as "a geographical area extending over two or more States determined by the watershed limits of the system of waters, including surface and underground waters, flowing into a common terminus."[589] Although this was the first time that any instrument issued by the ILA or IIL has addressed transboundary groundwater, the Comment on "Agreed Principles on International Law" issued by the ILA in 1958 highlighted the need to deal with groundwater. Article 1 stated that "A system of rivers and lakes in a drainage basin should be treated as an integrated whole." The Comment on this Article expressed the concern that international law had "for the most part been dealing with surface water, although there are some precedents having to do with underground waters. It may be necessary to consider the interdependence of all hydrological and demographic features of the drainage basin."[590]

The Helsinki Rules define a drainage basin to include both "surface and underground waters, flowing into a common terminus." It should be clarified that this definition includes only aquifers that are connected to surface water. Transboundary aquifers that do not contribute water to, or receive water from surface waters of an international drainage basin, are not covered by this definition.

Realizing this lacuna, the ILA issued the "Seoul Rules on International Groundwater in 1986" consisting of four articles.[591] The Rules use the terms "groundwater" and "aquifer" interchangeably, and the title of the first article is "the waters of international aquifers." The Rules define "aquifer" to include "all underground water bearing strata capable of yielding water on a practicable basis, whether these are in other instruments or contexts called by another name such as 'groundwater reservoir,' 'groundwater catchment area,' etc. including the

---

[589] Helsinki Rules, Article II; *supra* n. 166.

[590] *See* ILA, New York Report, *supra* n.163, at 99–100.

[591] *See supra* n. 297.

waters in fissured or fractured rock formations and the structures containing deep, so-called fossil waters."[592]

Article 1 defines international groundwaters as "the waters of an aquifer that is intersected by the boundary between two or more States…" and goes on to state that "…such an aquifer with its waters forms an international basin or part thereof." Furthermore, the Article characterizes those states as basin states within the meaning of the Helsinki Rules "…whether or not the aquifer and its waters form with surface waters part of a hydraulic system flowing into a common terminus." Article 2(1) deals with connected aquifers, and defines such aquifer as "An aquifer that contributes waters to, or receives water from, surface waters of an international basin … for the purposes of the Helsinki Rules."

Non-connected groundwater is dealt with in Article 2(2) of the Seoul Rules. This Article states that "An aquifer intersected by the boundary between two or more States that does not contribute water to, or receive water from, surface waters of international drainage basin constitutes an international drainage basin for the purpose of the Helsinki Rules."[593] Just as the Helsinki Rules pioneered the development of comprehensive international law rules relating to international rivers, the Seoul Rules set forth, for the first time, detailed rules related to all types of groundwater.

In addition, the Seoul Rules laid down more elaborate provisions requiring states, under Article 2(3), to take into consideration "any interdependence of the groundwater and other waters, including any interconnections between aquifers, and leaching into aquifers caused by activities in areas under their jurisdiction." Protection of shared groundwater from pollution is also emphasized and states are required, under Article 3, to prevent or abate such pollution in accordance with international law. The Article goes on to include obligations of states to consult and exchange information and data for the purpose of preserving groundwaters from degradation, and protecting, from impairment, the geologic structure of the aquifers, including recharge areas, as well as for the purpose of considering joint or parallel quality standards and environmental protection measures. Article 3(3) also includes the obligation to cooperate "…at the request of any one of them,…" in the collection and analysis of data pertaining to groundwaters or their aquifers. Article 4 recommends that the integrated management of shared groundwater resources, including conjunctive use with surface waters, should be considered by the states at the request of one of them. The accompanying report justifies the proposed articles on the following considerations:

First, the growing groundwater crisis, the legal implications of surface and underground waters interactions, and the characteristics of aquifers and their water have

---

[592] *See* footnote to Article 1 of the Seoul Rules, *id.*

[593] For the Helsinki Rules, *see supra* n. 166.

moved states generally to prescribe uncommon measures internally and, now, to call for analogous treatment for those transboundary aquifers already under stress.

Second, the number of international agreements expressly taking ground-waters into account is no longer negligible.

Third, state practice is demonstrating increasing willingness to accept the underground dimension of "transnational water resources."

Fourth, the UN International Law Commission in 1980 has expressly acknowledged in groundwater as a hydrographic component of an "international watercourse system."

As stated earlier, the ILA issued the Berlin Rules in 2004, consolidating and updating all its previous rules. Groundwater is dealt with in Chapter VIII of the Berlin Rules.[594] The Rules define the term "waters" to mean all surface water and groundwater other than marine waters. Chapter VIII applies to all types of aquifers, including aquifers that do not contribute water to, or receive water from, surface waters, or receive no significant contemporary recharge from any source, and calls for conjunctive, sustainable and precautionary management of aquifers. It defines a transboundary aquifer to include (i) an aquifer connected to surface waters that are part of an international drainage basin; or (ii) an aquifer intersected by the boundaries between two or more states even without a connection to surface waters that form an international drainage basin. The Chapter also calls for consultation, exchange of data and information, as well as joint management and protection of transboundary aquifers.[595]

Although the ILA made an explicit reference in the Seoul Rules to the work of the ILC in progress at that time on international watercourses, the Helsinki and Seoul Rules are, no doubt, the ones that have in the end influenced the work of the ILC on the UN Watercourses Convention, as discussed below. This was because they were the only Rules in place at that time.

8.2 The ILC Work on Transboundary Groundwater

In line with the ILA Helsinki Rules, the ILC defined the term "watercourse" in Article 2(a) of the draft UN Watercourses Convention, to consist of "a system of surface waters and groundwaters constituting by virtue of their physical rela-tionship a unitary whole and normally flowing into a common terminus." In

---

[594] *See* the Berlin Rules, *supra* n. 172. As indicated earlier, one major criticism of the Berlin Rules is the application of international principles to national waters. *See supra* n. 318. Chapter VIII of these Rules follows the same patterns and is intended to apply to national as well as transboundary aquifers.

[595] Chapter VIII of the Berlin Rules lays down in Article 41 detailed provisions on the pro-tection of transboundary aquifers, including from direct or indirect discharge of pollu-tants, injection of polluted water, saline water intrusion, and any other source of pollution.

this way the UN Watercourses Convention, like the Helsinki Rules, covers only aquifers that receive water from, or contribute water to, surface waters, and does not cover transboundary aquifers that are not connected to surface waters.[596]

According to this definition, the two components of the system should not only constitute a unitary whole, but they should also normally flow "into a common terminus." Those two criteria for the relationship between surface waters and groundwaters makes it clear that groundwater not connected to surface water is not covered by the provisions of the Convention. However, after the ILC completed its work on the draft Watercourses Convention, it decided to address the issue of non-connected groundwater that was not included in the draft Convention. This situation resembles the one in which the ILA found itself after it passed the Helsinki Rules. Just like the Helsinki Rules were complemented by the Seoul Rules in 1986, the ILC adopted in 1994 the "Resolution on Confined Transboundary Groundwater" to complement the Watercourses Convention.[597]

The Resolution defined confined groundwater as "groundwater not related to international watercourses,"[598] and recognized that such groundwater is also a natural resource of vital importance for sustaining life, health and the integrity of ecosystems. The Resolution also recognized the need for continuing efforts to elaborate rules pertaining to confined transboundary groundwater, and makes the following three recommendations:

First, states shall be guided by the principles contained in the draft articles on the Law of the Non-navigational Uses of International Watercourses, where appropriate, in regulating transboundary groundwater.

Second, states should consider entering into agreements with the other state or states in which the confined transboundary groundwater is located.

Third, in the event of any dispute involving transboundary confined groundwater, the states concerned should consider resolving such dispute in accordance with the provisions contained in Article 33 of the draft articles of the Watercourses Convention, or in such other manner as may be agreed upon.

---

[596] For a general discussion of the matter, see Gabriel Eckstein, *A Hydrogeological Perspective of the Status of Ground Water Resources Under the UN Watercourse Convention*, 30 Columbia Journal of Environmental Law 525 (2005).

[597] For the International Law Commission Resolution, see Yearbook of the International Law Commission, 1994, Vol. 2, at 135 (1997). For discussion of the issue, see Stephen McCaffrey, *International Groundwater Law: Evolution and Context*, in *Groundwater: Legal and Policy Perspectives, supra* n. 368, at 139; *see* also Raj Krishna & Salman M. A. Salman, *International Groundwater Law and the World Bank Policy for Projects on Transboundary Groundwater, id.,* at 163.

[598] The definition of the term "confined groundwater" in this sense has been criticized as incorrect and not scientific, and the use of the term with this definition has been discontinued. For discussion of the term and what it entails, *see* Eckstein, *supra* n. 596, and *infra* n. 601.

As such, the Helsinki, Seoul and Berlin Rules, as well as the UN Watercourses Convention and the ILC Resolution have made considerable contribution to the emergence and development of international law rules for the use and protection of transboundary groundwater.

The United Nations General Assembly (UNGA) returned to the topic of transboundary groundwater in 2002, five years after it adopted the Watercourses Convention. In that year, the UNGA took note of the decision of the ILC to proceed with its work on a number of topics, including "Shared Natural Resources."[599] The ILC agreed that the topic covers groundwaters, and oil and natural gas, but decided to adopt a step-by-step approach, and to focus on the consideration of transboundary groundwaters as a follow-up to the ILC previous work on the Watercourses Convention. As a result, it also decided to embark first on the codification of the law of transboundary aquifers independently from any future work on oil and natural gas.[600]

The ILC Special Rapporteur issued five reports between 2003 and 2008 on transboundary aquifers. In 2006, the ILC adopted, on first reading, the Draft Articles on the Law of Transboundary Aquifers, consisting of 19 articles, based largely on the UN Watercourses Convention.[601] The Draft Articles were transmitted to Governments for comments and observations, with a request that such comments be submitted by January 1, 2008. While awaiting the comments from Governments, the ILC addressed the question of the relationship between its work on transboundary aquifers and that on oil and natural gas. It decided to proceed with and complete the second-reading of the Law of Transboundary Aquifers independently of its possible future work on oil and natural gas. The ILC received written and oral comments from 47 states which were generally favorable and supportive of the ILC work. Those comments were considered during the second reading, and the revised Draft Articles were finally adopted by the ILC on August 5, 2008.[602] As will be discussed below, the UNGA adopted a resolution

---

[599] UNGA Resolution 57/21, Report of the International Law Commission on the Work of its Fifty-Fourth Session, November 19, 2002. *See* also Report of the Fifty-Fourth Session of the International Law Commission (April 29 to June 7, and July 22 to August 17, 2002), available at: http://untreaty.un.org/ilc/sessions/54/54sess.htm.

[600] *See* Report of the Fifty-Fifth Session of the International Law Commission (May 5 to June 6 and July 7 to August 8, 2003), available at: http://untreaty.un.org/ilc/sessions/55/55sess.htm.

[601] For an analysis of the Draft Articles, *see* Gabriel E, Eckstein, *Commentary on the U.N. International Law Commission's Draft Articles on the Law of Transboundary Aquifers*, 18 Colo. J. Int'l Envt. L & Pol'y, 537 (2007).

[602] *See* Report of the International Law Commission at its Sixtieth Session (May 5 to June 6, and July 7 to August 8, 2008); available at: http://daccessdds.un.org/doc/UNDOC/GEN/N08/249/11/PDF/N0824911.pdf?OpenElement.

taking note of the Articles on the Law of Transboundary Aquifers (the Articles) four months later.

The Articles follow, by and large, the principles enunciated in the UN Watercourses Convention with adaptations to fit the special characteristics of aquifers. Article 2 defines an aquifer as "a permeable water-bearing geological formation underlain by a less permeable layer and the water contained in the saturated zone of the formation," and "aquifer system" to mean a series of two or more aquifers that are hydraulically connected. The terms "transboundary aquifer" and "transboundary aquifer system" are defined as "an aquifer or aquifer system, parts of which are situated in different states."[603]

Article 4 calls on the states to utilize the transboundary aquifers according to the principle of equitable and reasonable utilization, keeping in mind the need to maximize the long-term benefits, to establish comprehensive utilization and sustainability plans of the aquifer. Article 5 adopts and adapts the factors for equitable and reasonable utilization included in the Watercourses Convention, while Article 6 deals with the obligation not to cause significant harm. The Articles also include provisions on the general obligation to cooperate, and on exchange of data and information. Furthermore, Article 9 encourages the aquifer states to enter into bilateral or regional agreements or arrangements with respect to the entire aquifer or aquifer system, or any part thereof, or a particular project or program.

Article 10 deals with protection and preservation of ecosystems within or dependent upon the transboundary aquifer,[604] and Article 11 requires the aquifer states to take appropriate measures to prevent and minimize detrimental impacts on the recharge and discharge processes. Article 12 calls for prevention, reduction

---

[603] This definition is limited to groundwater only, and does not take into account the interconnections between groundwater and surface water. As discussed earlier, the UN Watercourses Convention deals with the connected system of surface waters and groundwaters. According to Julio Barberis, groundwater may become of international relevance in the following situations: (i) where a confined aquifer is intersected by an international boundary; (ii) where an aquifer lies entirely within the territory of one state but has interconnections and interdependence with an international watercourse; (iii) where the aquifer is entirely situated within the territory of one state but has interconnections and interdependencies with another aquifer in another state; and (iv) where an aquifer is entirely situated within the territory of one state but is getting recharged in another state. *See* Julio A. Barberis, *International Groundwater Resources Law,* FAO Legislative Study No. 40, at 36 (FAO 1986). One situation that may not be covered by these definitions relates to a river that runs exclusively in one state, but has connections to an international aquifer. Would such a river be considered as an international watercourse? Based on the definition of the term "watercourse" under the UN Watercourses Convention (*see supra* n. 267), such a river would be considered as an international watercourse.

[604] The Articles use the concept of the ecosystem, which is broad and includes flora, fauna and land contiguous to the water resources. For further analysis of the concept of "ecosystem" *see* David Hunter, James Salzman & Durwood Zaelke, *International Environmental Law and Policy* at 482 (Foundation Press 1998).

and control of pollution, and urges adoption of a precautionary approach in view of uncertainty about the nature and extent of a transboundary aquifer or aquifer system and of its vulnerability to pollution.[605]

Notification for planned activities is addressed in Article 15, and is based largely on the corresponding articles of the Watercourses Convention. It requires the aquifer state that plans to implement any activities which may affect a transboundary aquifer or aquifer system, and thereby may have significant adverse effect upon another aquifer state, to provide that state with timely notification thereof. Such notification shall be accompanied by available technical data and information, including any environmental impact assessment, in order to enable the notified state to evaluate the possible effects of the planned activities. Article 16 deals with technical cooperation with developing states, while Article 17 addresses emergency situations. Protection of the transboundary aquifers or aquifer systems in time of armed conflict is dealt with in Article 18, while Article 19 addresses the issue of data and information vital to national defence or security, requiring states to cooperate in good faith with a view of providing as much information as possible under the circumstances.

The Special Rapporteur noted that the views of governments varied from supporting a legally binding convention to a non-binding document. Accordingly, he decided to recommend that the General Assembly (a) take note of the Draft Articles on the Law of Transboundary Aquifers in a resolution, and annex the Draft Articles to the resolution; (b) recommend that states make appropriate arrangements bilaterally or regionally with the states concerned for proper management of their transboundary aquifers on the basis of the principles enunciated in these articles; and (c) also consider, at a later stage, and in view of the importance of the topic, the possibility of convening a negotiating conference to examine the draft articles with a view to concluding a convention.[606]

Since some time would elapse before a decision is made on the second step of elaboration of a convention, the ILC did not include an article on the relation of the Draft Articles to other conventions and international agreements. Nor did the ILC include provisions on dispute settlement. However, an article on the relation to other conventions was proposed and included in the Report.[607] With regard to dispute settlement, it was suggested to include provisions thereon if the text of the Articles should take the form of a convention.[608]

---

[605] The Articles use the term "precautionary approach," rather than the term "precautionary principle" similar to the Berlin Rules (*supra* n. 595). For more discussion of the topic, *see The Precautionary Principle and International Law: The Challenge of Implementation* (David Freestone & Ellen Hey, eds., Kluwer Law International 1996).

[606] *See* paragraph 9 of the ILC Report, *supra* n. 602.

[607] *See* paragraph 38 of the ILC Report, *supra* n. 602.

[608] *See* paragraph 41 of the ILC Report, *supra* n. 602. *See* also Salman M. A. Salman, *The International Law Commission Adopts Draft Articles on the Law of Transboundary*

The ILC recommendation of a two-step approach to the UN General Assembly, consisting of adopting the Articles now, and considering elaboration of a convention at a later stage, is realistic, given the delays and difficulties facing the entry into force of the Watercourses Convention itself.[609] This approach is similar to that adopted with regard to the Draft Articles on Responsibility of States for International Wrongful Acts. As recommended by ILC then, the General Assembly took note of the articles on responsibility of states for internationally wrongful acts, the text of which was annexed to the UNGA Resolution, and recommended them to the attention of Governments without prejudice to the question of their future adoption or other appropriate action.[610]

During its meeting on December 11, 2008, the UNGA welcomed the conclusion of the work of the ILC on the law of transboundary aquifers and its adoption of the Articles, together with a detailed commentary, on the subject.[611] It expressed appreciation to the ILC for its continuing contribution to the codification and progressive development of international law, and to the International Hydrological Programme of UNESCO and to other relevant organizations for the valuable scientific and technical assistance rendered to the ILC. Furthermore, the UNGA took note of the Articles on the Law of Transboundary Aquifers, presented by the ILC, which are annexed to the UNGA Resolution, and commended them to the attention of governments without prejudice to the question of their future adoption or other appropriate action, and encouraged the States concerned to make appropriate bilateral or regional arrangements for the proper management of their transboundary aquifers, taking into account the provisions of these Articles. The Resolution indicated that the UNGA will include in the provisional agenda of its sixty-sixth session an item entitled "the Law of Transboundary Aquifers" with a view to examining, *inter alia*, the question of the form that might be given to the Articles.

The adoption by the UNGA of the resolution taking note of the Articles on the Law of Transboundary Aquifers concluded an important stage of the work on the

---

Aquifers, Issue 7, the Nature of Law (September 2008), available at:http://newsletters. worldbank.org/external/default/main?menuPK=2363052&theSitePK=2363040&page PK=64133601&content MDK=21905856&piPK=64129599.

[609] *See* Salman, *The United Nations Watercourses Convention Ten Years Later, supra* n. 181, at 10.

[610] *See* UNGA Resolution 56/83 of December 12, 2001. The General Assembly issued Resolution 62/61 on January 8, 2008, commending once again the Articles on Responsibility of States for Internationally Wrongful Acts to the attention of governments, without prejudice to the question of their future adoption or other appropriate action.

[611] UNGA Resolution A/RES/63/124. *See* Report on the Sixty-Third Session of the UNGA, available at: http://www.un.org/News/Press/docs/2008/ga10798.doc.htm. This Resolution was adopted by the UGA without a vote.

topic that commenced in 2002, and lasted for six years. No doubt this is a landmark step in the codification and progressive development of the law of the international watercourses, covering a major gap in the UN Watercourses Convention, and extending application of the basic principles on international water law to all types of groundwater.

8.3 Transboundary Groundwater under International Agreements

Recognition of transboundary groundwater in bilateral and multi-lateral treaties evolved slowly before the 1990s, but accelerated considerably thereafter.[612] As early as 1964, the Convention and Statutes of Lake Chad Basin made an explicit reference to groundwater when they stated that the exploitation of the Basin "…and especially the utilization of surface and underground waters has the widest meaning and refers in particular to the needs of domestic, industrial and agricultural development…"[613] The Statutes, as indicated earlier, oblige the member states of the Lake Chad Basin Commission (LCBC) to notify and consult with the LCBC before undertaking any work related to the development of the water resources of the Basin.

Along the same lines, and as discussed earlier, Article V of the African Convention on the Conservation of Nature and Natural Resources, concluded in 1968, deals also with groundwater. It requires the contracting states to consult with each other, and if the need arises, to set up inter-state commissions to study and resolve problems arising from the joint use of surface and underground water resources, and for their joint development and conservation.[614] Joint management is also addressed by the Arrangement on the Protection, Utilisation, and Recharge of the Franco-Swiss Genevois Aquifer between the State Council of the Republic

---

[612] *See* Food and Agriculture Organization, *Groundwater in International Law—Compilation of Treaties and other Legal Instruments*, FAO Legislative Study 86 (2005).

[613] *See* Article 4 of the Statutes of Lake Chad Basin, *supra* n. 137. Although the 1950 Agreement between Germany and Luxembourg dealt with construction of a dam by Luxembourg on the Sauer River, the Agreement addressed the effects of the dam on groundwater. The Agreement stated that "in the event of damage caused by a rise or fall in the ground water on the west side of the Sauer in consequence of the construction of the dam, the government of the Grand Dutchy of Luxembourg undertakes to rectify such damage or pay appropriate compensation." *See* State Treaty Between the Grand Duchy of Luxembourg and the Land Rhineland-Palatinate in the Federal Republic of Germany Concerning the Construction of a Hydro-electric Power-plant on the Sauer (Sûre) at Rosport/Ralingen, Apr. 25, 1950, F.R.G.-Lux., art. 10, *reprinted in* Legislative Text and Treaty Provisions Concerning the Utilization of International Rivers for Other Purpose than Navigation, U.N. Doc. ST/LEG/SER.B/12 (1963) at 721–723.

[614] *See supra* n. 361.

and Canton of Geneva and the Prefect of Haute-Savoie, concluded in 1977. The Arrangement laid down detailed provisions for the joint management of the Aquifer, including the establishment of a commission for that purpose.[615]

The 1980 Convention Creating the Niger Basin Authority (NBA) is one of the early instruments where the interconnections between surface and groundwater were recognized. The Convention spells out the objectives of the NBA which include "the initiating and monitoring of an orderly and rational regional policy for the utilization of the surface and underground waters in the Niger Basin." [616] Thus, under the Convention a regional, and not just a country specific, policy would be initiated and monitored by the NBA itself.

The 1991 Espoo Convention makes a more explicit reference to notification regarding transboundary groundwater.[617] As mentioned before, the Convention requires notification by the Party of Origin of any proposed activity listed in Appendix I to the Convention, which is likely to cause another party significant adverse transboundary impact. The list includes "groundwater abstraction activities in cases where the annual volume of water to be abstracted amounts to 10 million cubic meters."[618] The Helsinki Convention defines "transboundary waters" to include "any surface or groundwaters which mark, cross, or are located on boundaries between two or more states. . . ."[619] As discussed earlier, the Convention requires the parties to provide for the widest exchange of information, and to consult as early as possible, on issues covered under its provisions.[620]

---

[615] For a copy of an unofficial translation of the Arrangement *see*: http://www.waterlaw. org/regionaldocs/franko-swiss-aquifer.html. The Arrangement was updated in 2008 *see*: http://www.waterlaw.org/regionaldocs/2008Franko-Swiss-Aquifer-English.pdf.

[616] *See* Article 1(d) of the Convention, *supra* n. 411.

[617] *See supra* n. 366. Reference in this connection should also be made to the Charter on Groundwater Management adopted by the UN/ECE at its forty-fourth session in 1989 (E/ECE/1197 ECE/ENVWA/12), available at: http://www.internationalwaterlaw.org/ documents/regionaldocs/groundwater_charter.html. Paragraph XXV of the Charter deals with "International Cooperation" and sets forth a number of obligations on the riparian countries, including "the obligation to give notification concerning any activity which might modify the volume and/or the quality of groundwater."

[618] *See* paragraph 12 of Appendix I of the Convention (List of Activities). As discussed earlier, the European Bank for Reconstruction and Development (EBRD) incorporated this requirement of the Espoo Convention in "EBRD Environmental and Social Policy." Those types of projects, together with a number others, are classified as Category A projects requiring "formalized and participatory assessment process carried out by third party specialist." However, no requirement of notification of other riparians is specified under the EBRD policy, perhaps because the notification is already required under the Espoo Convention, *see supra* n, 206.

[619] *See* Article 1(1) of the Helsinki Convention; *supra* n. 365.

[620] *See id.,* Article 6.

In addition, both the 1994 Danube Convention, and the 1999 Convention for the Protection of the Rhine extend their application to groundwater in the Basin of each of the Danube and Rhine, respectively.[621]

The 2000 SADC Protocol defines the term "watercourse" to mean a system of surface waters and groundwaters constituting by virtue of their physical relationship a unitary whole normally flowing into a common terminus. This definition is based on that of the UN Watercourses Convention. Similarly, the Protocol requires the parties to exchange information and consult each other and, if necessary, negotiate the possible effects of planned measures on the condition of a shared watercourse. Similar to the UN Watercourses Convention, the Protocol also requires notification of other riparian states of any planned measures which may have significant adverse effect on any of such riparians.[622]

A number of recent agreements in Africa define the term "basin" to include both surface water and groundwater. Those instruments include provisions on the exchange of data and information, and for notification. The 2003 Lake Victoria Protocol[623] defines the Lake Victoria Basin as the "geographical area extending within the territories of the Partner States determined by the watershed limits of the system of waters, including surface and underground waters flowing into Lake Victoria."[624] The Protocol also obliges each of the Partner States to notify the other Partner States of planned activities within its territory that may have adverse affects upon those other States.[625] Similar provisions regarding the definition of the watercourse (to include both surface and ground water), and the requirement of notification of other riparians, are included in the

---

[621] The Danube Convention (*see supra* n. 215) states that the Contracting Parties shall strive to achieve the goals of a sustainable and equitable water management, including the conservation, improvement and the rational use of surface waters and ground water in the catchment area as far as possible. The Convention on the Protection of the Rhine (also referred to as the 1999 Rhine Convention) states that the Convention applies to: the Rhine; ground water interacting with the Rhine; and aquatic and terrestrial ecosystems which interact or could again interact with the Rhine. For the 1999 Rhine Convention, *see*: http://www.iksr.org/index.php?id=327. Furthermore, the Framework Agreement on the Sava River Basin (*supra* n. 215) states in Article 1 that the Sava River Basin "is the geographical area extended over the territories of the Parties, determined by the watershed limits of the Sava River and its tributaries, which comprises surface and ground waters, flowing into a common terminus."

[622] *See* Article 4 of the SADC Protocol, *supra* n. 56.

[623] *See* Protocol for Sustainable Development of Lake Victoria Basin, Article 1, *supra* n. 447.

[624] *See id.*, Article 1(2).

[625] *See id.,* Article 4(2) (d).

2003 Lake Tanganyika Convention,[626] and the Agreement on the Establishment of the Zambezi Watercourse Commission,[627] as well as the 2002 Water Charter of the Senegal River.[628]

In connection with the emerging international law on groundwater, mention must also be made of the Bellagio Draft Treaty which is the work of a group of multidisciplinary specialists in the field of transboundary groundwater. It is proposed to serve as a model international groundwater treaty between the states sharing the aquifer. The Draft Treaty "is based on the proposition that water rights should be determined by mutual agreement rather than be the subject of uncontrolled, unilateral taking, and that rational conservation and protection actions require joint resource management machinery."[629] The overriding goal of the Bellagio Draft Treaty is "to achieve joint, optimum utilization of the available waters, facilitated by procedures for avoidance or resolution of differences over shared groundwaters in the face of the ever increasing pressures on this priceless resource."[630] The Bellagio Draft Treaty consists of 20 articles and includes detailed provisions for specific cooperative arrangements required for the proper management, exploration and development of shared groundwater resources.

---

[626] *See* The Convention on the Sustainable Management of Lake Tanganyika, concluded on June 20, 2003, by Burundi, Democratic Republic of Congo, Tanzania and Zambia. The definition of the Lake Tanganyika Basin is in fact more inclusive than that of Lake Victoria. Article 1 defines the Basin to include "the whole or any component of the aquatic environment of Lake Tanganyika and those ecosystems and aspects of the environment that are associated with, affect or are dependent on, the aquatic environment of Lake Tanganyika, including the system of surface and ground waters that flow into the Lake from the Contracting States and the land submerged by these waters." Article 14 of the Convention deals with the obligation of prior notification. (Copy of the Convention on file with author.)

[627] The Agreement on the Establishment of the Zambezi Watercourse Commission was concluded on July 13, 2004 by Angola, Botswana, Malawi, Mozambique, Namibia, Tanzania, and Zimbabwe. Zambia is not yet a signatory to the Agreement, *see supra* n. 267. The Agreement defines the Zambezi Watercourse to include "the system of surface and ground waters of the Zambezi constituting by virtue of their physical relationship a unitary whole flowing normally into common terminus, the Indian Ocean." Article 16 obliges the parties to notify the Zambezi Commission of any planned activities which may adversely affect the Zambezi Watercourse. (Copy of the Agreement on file with author.)

[628] *See supra* n. 392. Article 1(20) of the Charter defines groundwater to include water contained in porous, permeable and/or fissured geological formations totally and or partially renewed by the hydrological flow of the Senegal River. Article 4 obliges the parties to inform the other riparian states before engaging in any activity or project likely to have an impact on water availability.

[629] *See* Robert D. Hayton & Albert E. Utton, *Transboundary Groundwaters: The Bellagio Draft Treaty*, 29 Nat. Resources J. 663 (1989). For the text of the draft treaty, *see* Salman, *supra* n. 368, at 221.

[630] *See* Hayton & Utton, *id.,* at 665.

This overview of the work of the ILA and ILC, as well as the global, regional and multi-lateral agreements on international watercourses indicates a clear and strengthening trend towards defining shared surface and ground water as the whole or a component of the aquatic environment. The overview also shows an accelerated movement towards recognizing the inter-connections between surface and ground water, and the establishment of the same requirements for exchange of data and information, as well as notification, for both of them.

8.4   World Bank Approach to Transboundary Groundwater

As discussed in Chapter 4 of this book, the Bank policy for projects on international waterways refers explicitly to surface waters and semi-enclosed coastal waters only. The idea of extending application of the policy to transboundary groundwater was considered in 1984. The committee that prepared the Memorandum to the Executive Directors[631] justified the non-inclusion of groundwater by the lack of adequate scientific knowledge. The committee also wanted to wait for the completion of the work of the ILC that had not, by that time, agreed on how to address transboundary groundwater.[632]

The Bank faced the challenges of transboundary groundwater for the first time in 1990, five years after it adopted the 1985 OMS. Prior to 1990, the Bank had financed a number of groundwater projects in some countries including Yemen, Tunisia, Jordan, Mexico, Nepal and Bangladesh, but "....the complexity of the issues of shared groundwater resources did not arise presumably because either the possibility of whether the concerned aquifers might be shared with neighboring States was not entertained, or simply because it was assumed the aquifers concerned were purely national."[633]

The Sahara Regional Development Project in Algeria was the first project in which the policy on international waterways was extended to transboundary groundwater.[634] The Project involved the utilization of waters by Algeria from two aquifers through the drilling of wells and provision of related equipment and installations, and closure of deteriorating and non-operational wells. One of the aquifers, known as the "Continental Interclaire" (CI) holds fossil, non-renewable waters. The other aquifer which is known as the "Complex Terminal" (CT) is a rechargeable aquifer. While it has been established by geological studies that

---

[631] *See* Memorandum to the Executive Directors, *supra* n. 151.

[632] *See* Krishna, *supra* n. 105, at 40.

[633] *See* Krishna & Salman, *supra* n. 597, at 180.

[634] *See* Algeria, Sahara Regional Development Project (P004938, 1991).

these aquifers extend from Algeria to Tunisia, the possibility of their extending further east to Libya and west to Morocco has been indicated by diagrams in some reports.[635] An area of about 800,000 cubic kilometers of these aquifers has been studied by the United Nations Educational, Scientific and Cultural Organization (UNESCO) during the period 1968–1971, with funding from the United Nations Development Programme (UNDP), as part of a strategy to explore water resources in the desert areas of the world. The studies were undertaken with the intention of assessing the water resources potential in these areas and to propose a policy for their exploitation by both Algeria and Tunisia. In 1983 UNDP reviewed and updated this study.[636]

As part of the policy of exploitation, three requirements were proposed in the studies as follows: (i) the artesianism of the CI should be maintained until 2010; (ii) the pumping of the CT should not exceed 60 meters until 2010; and (iii) extraction by each country should have the same impact on the other. Both the Algerian and Tunisian sides appear to have endorsed the study and recommendations in 1989.

The Project documents issued in May 1990 mentioned, for the first time, the possibility of "riparian issues" in the proposal that the Bank should seek commitment from the Government of Algeria to adhere to the recommendations of the UNDP study for the use of the common groundwater resources. The deliberations within the Bank raised the facts that (i) OP 7.50 does not expressly cover groundwater, (ii) the UNDP study would ensure the best use of the common resources and protection of the common international interests, and (iii) the *"Commission Technique Tuniso/Algerienne pour l'Hydraulique et l'Environnement"* (CTHE) should serve as the appropriate forum for the discussion of the issues raised by the proposed project.

However, at that point the issue of applying OD 7.50 to the Project started to come under serious consideration, based on the fact that the water in question was indeed shared by two states. After more debate, it was decided that Algeria should notify Tunisia of the Project, and that such notification should: (i) indicate the Algerian Government's endorsement of the UNDP study, and understanding of a similar endorsement on the part of the Tunisian Government; (ii) emphasize the measures the Algerian Government plans to take to prevent pollution of the aquifers, and its expectation for similar measures on the part of the Tunisian Government; (iii) request comments on the Project from the Tunisian side; and (iv) give Tunisia a period of not less than one month to respond.

Since the extension of the aquifers to Libya and Morocco was not supported by actual geological studies at that time, and since ascertaining such an extension

---

[635] For an update on this issue, *see infra* n. 637.

[636] An updated version of the study is available at: http://nwsas.iwlearn.org/.

would have been very difficult and costly, notifying these two countries of the proposed project was considered unwarranted.[637] The possibility of the World Bank notifying Tunisia in the event of a request for such an action by Algeria was also considered in case Algeria asked the Bank to do so on its behalf.

After corresponding with Algeria, the Bank, on October 24, 1990, notified the Tunisian authorities of the project, provided them a summary of the Project and its impact on the shared aquifer, as well as the Project Evaluation Report. The notification emphasized the Bank's understanding of the existence of cooperative arrangements between the two countries regarding the aquifers, and their intention to adhere to the UNDP study recommendations. Tunisia was given one month to respond.

No response was received from Tunisia. The Bank proceeded with the processing of the Project, and the project was presented to the Executive Directors of the Bank on September 24, 1991. It is worth adding that the legal documents included a covenant stating that ". . .the Borrower shall take all measures necessary to ensure that the use of the groundwater resources in [the Northern Sahara] Region shall be planned in accordance with the recommendation of the Study of Water Resources in the Sahara."[638]

---

[637] The two aquifers are now treated as one aquifer, and it has been established since the mid-1990s that Libya also shares this Aquifer with Algeria and Tunisia. The Aquifer is now called the North Western Sahara Aquifer System (NWSAS), although it is commonly referred to as SASS, a French acronym for "*Système Aquifère du Sahara Septentrional.*" The Aquifer covers an area of over one million square kilometers; 700,000 in Algeria, 250,000 in Libya, and 80,000 in Tunisia. *See*: http://www.oss-online.org/pdf/synth-sass _En.pdf. The Minutes of a meeting held in Rome on December 19 and 20, 2002 recorded the agreement of the three parties for the Establishment of a Consultation Mechanism for the Aquifer. The objective of the Consultation Mechanism is to coordinate, promote and facilitate the rational management of the Aquifer. For excerpts from the Minutes *see* FAO, *Groundwater in International Law supra* n. 613, at 6. The Consultation Mechanism was inaugurated in June 2008, under the auspices of *Observatoire du Sahara et Sahel* (OSS), following the signature of the Declaration by the Ministers of Water Resources of Algeria, Libya and Tunisia for the North-Western Sahara Aquifer System. The Declaration states that the main objective of the Mechanism is to serve as a framework for cooperation and exchange of information among the three riparian states. Since its establishment, the Mechanism has held a series of workshops and meetings in the three countries sharing the Aquifer. *See*: http://www.oss-online.org/pdf/synth-sass_En.pdf. The OSS is an independent international organization based in Tunis, Tunisia. It was founded in 1992 to improve early warning and monitoring systems for agriculture, food security and drought in Africa. Membership of OSS consists of 22 African countries, five countries in Europe and North America (Germany, Canada, France, Italy and Switzerland), and four sub-regional African organizations. For more information on OSS, *see*: http://www.oss-online.org/index.php? option=com_content&task=view&id=433&Itemid=564& lang=en.

[638] Section 4.02 of the Loan Agreement between Democratic and Popular Republic of Algeria and International Bank for Reconstruction and Development, dated November 6, 1992 (Loan No. 3405-AL).

Thus, the Bank policy for projects on international waterways which refers only to surface water was extended for the first time to transboundary groundwater. Accordingly, a precedent was established in 1990. The precedent, no doubt, rests on solid legal basis. The rationale for notification for surface water applies equally, and arguably more strongly, to groundwater. The Bank has been acting in the best interests of all riparians in case of surface water, and needs to act the same way in the case of groundwater. As the aforementioned discussion on the international instruments has shown, since 1990 there has been considerable recognition in many regions of the need for collaborative arrangements for shared groundwater. Moreover, and as discussed in the previous Part of this Chapter, a large number of regional and multilateral treaties have been concluded defining the term "watercourse" to include both surface and groundwater, and requiring exchange of date and information and notification for shared groundwater. The Bank cannot possibly ignore these treaties. As discussed earlier, the Bank cannot finance a project that contravenes the borrower's obligations under an international environmental treaty.[639]

Indeed, the Bank acknowledged in 1992 the international character and complexities of shared aquifers. The World Bank Policy Paper states that "International water issues also involve important groundwater resources. In a number of cases, aquifers cross international boundaries; thus pumping by one country interferes with another country's pumping or stream flows."[640]

It is worth noting that it was again the Middle East which prompted another milestone in the evolution and development of the Bank policy for projects on international waterways. It may be recalled that it was the Ghab Project in Syria, and Turkey's objection that resulted in the issuance of the first Bank policy in 1956. Subsequently, it was the Igdir-Aksu Project in Turkey, and the objection by Iran that prompted the Bank to issue a new elaborate policy in 1985. By establishing the basis for adding groundwater to the policy, the Algeria project, no doubt, is another landmark in the evolution of the policy.

OD 7.50 which was issued in 1990, replacing OMS 2.32 of 1985, was converted in 1994 into Operational Policy (OP), and Bank Procedures (BPs), in line with the conversion of the other Operational Directives of the Bank into this new architecture. The issue of expanding the definition of international waterways under the policy to include transboundary groundwater was considered, but groundwater was not included, perhaps because the 1994 change

---

[639] *See supra* nn. 553 & 554.

[640] *See* The World Bank, *Water Resources Management Policy Paper* 1993, *supra* n. 233, at 39.

was considered a mere conversion of the Directive into OP/BP, and not a revision of the policy.[641]

The Bank faced the issue of transboundary groundwater again in 1997, a few years after the Sahara Regional Development Project. This time it was the Disi-Amman Water Conveyor Project, and again it was the Middle East region. Under the proposed Project, Jordan would convey water from the Disi Aquifer to Amman through a 350-kilometer conveyor. A build, operate, transfer (BOT) contract was proposed for which the Bank was expected to offer to the bidders a partial risk guarantee that would protect commercial lenders against government default.[642] The Disi Aquifer is a fossil, non-renewable aquifer which underlies both northern Saudi Arabia and southern Jordan. However, no joint management mechanism for the aquifer existed between the two countries at that time. As Greg Shapland observed, "There is a perceptible flow from the Saudi side into Jordan, with water from the aquifer eventually emerging at the surface near the Dead Sea. The reserves are large, but, because much of it is more than 250 meters below ground (the conventional economic pumping limit), it may not be cost effective to abstract more than a small proportion."[643] The proposed Project would abstract up to 100 million cubic meters a year (MCM/yr.), with the likelihood that the figure may climb to 150 MCM/yr. in the future to meet the growing demand for water in Amman. The well fields to be drilled would be about 35 kilometers from the borders with Saudi Arabia.

Although the Bank took note of the fact that OP/BP 7.50 refers only to surface water, the precedent of the Sahara Regional Development Project in Algeria was considered. It was decided that the precedent should be followed. Accordingly, the Bank informed Jordan of the need to notify Saudi Arabia of the Project. The Risk Assessment indicated that expected abstractions in Saudi Arabia would not significantly affect water levels in the Jordanian well fields, but that any new and unexpected developments close to the border represented a potential project risk.[644] Jordan agreed and notified Saudi Arabia of the Project in February 1998.

---

[641] As discussed earlier, the 1994 OP included for the first time a third exception to the notification requirement under paragraph 7 of OP 7.50. That exception relates to a project in a tributary of an international waterway where the tributary runs exclusively in the lowest downstream riparian. *See supra* Chapter 7, Part 7.2.3 of this Book. As mentioned earlier, this was the only change from the previous OD introduced into the 1994 OP.

[642] For a detailed discussion of the Project, *see* Andrew Macoun & Hazim El Naser, *Groundwater Resources Management in Jordan: Policy and Regulatory Issues*, in *Groundwater, supra* n. 368, at 105.

[643] *See* Shapland, *supra* n. 47, at 148.

[644] *See supra* n. 642, at 112.

Saudi Arabia requested more information, and later gave a qualified no objection.[645] However, processing of the project was discontinued by the Bank because of some issues concerning the partial risk guarantee.

This project followed and strengthened the precedent of inclusion of transboundary groundwater in the scope of international waters covered under the policy, and of requiring notification of other riparians of projects that may affect the shared aquifer.[646]

Subsequent to these two projects, a number of projects involving the use of shared aquifers were financed by the Bank, and notification of other riparians was undertaken. One such project was the 2003 Second National Fadama Development Project in Nigeria.[647] The project consisted of financing demand driven micro-projects which would include small scale irrigation and water supply projects, using both surface and groundwater in the Niger and Lake Chad basins. Nigeria notified all the riparian states as well as the Lake Chad Basin Commission (LCBC), and the Niger Basin Authority (NBA). No concerns were expressed by any of the riparians, the LCBC or the NBA. A similar approach was followed by Nigeria for the Fadama III project in 2007.[648] Unlike the previous project, Fadama III involved only the abstraction of groundwater in the Niger Basin catchment area, and no surface water was involved. The NBA was notified by Nigeria and it gave a positive response.

Similarly, three projects in Ethiopia in 2004 involved the use of both surface and groundwaters shared with other countries.[649] The surface waters under the

---

[645] Saudi Arabia asked that no wells should be drilled within 15 kilometers of the borders in each country. *See supra* n. 478.

[646] These two projects have generated discussion within the Bank as to whether the Bank should be financing projects that affect fossil aquifers. The World Bank Operational Policy on Water Resources Management (OP 4.07) states in paragraph 2(d) that the Bank assists borrowers in "restoring and preserving aquatic systems and guarding against overexploitation of groundwater resources, giving priority to the provision of adequate water and sanitation services for the poor." One important factor to be considered regarding the question of whether to finance projects that affect fossil aquifers is the project alternatives. Paragraph 2 of OP 4.01, Environmental Assessment (EA) states that: "EA is a process whose breadth, depth, and type of analysis depend on the nature, scale, and potential environmental impact of the proposed project. EA evaluates a project's potential environmental risks and impacts in its area of influence; examines project alternatives ..." The analysis of alternatives compares feasible alternatives to the proposed project site, technology, design, and operation—including the "without project" situation. *See* Annex B to OP 4.01, *see infra* n. 695.

[647] Nigeria, Second National Fadama Development Project (P063622, 2003).

[648] Nigeria, Third National Fadama Development Project (Fadama III) (P096572, 2008).

[649] Those projects are (i) Water Supply and Sanitation Project, *supra* n. 438, (ii) Decentralized Infrastructure project, *supra* n. 438, and (iii) Productive Safety Net Program, *supra* n. 468.

project related to the Nile and the Wabe Shabelle and Ogaden basins, while the groundwater related to the Rift Valley Lakes. All the Nile riparian countries, as well as Somalia and Djibouti, were notified of the project, and no unfavorable responses were received from any of them. This trend of dealing with shared surface and groundwater continued and the Inland Waters Project in Croatia in 2007 followed the same approach. Under the project, new well fields would be developed in the Danube River catchment area within Croatia. Although groundwater extraction under the project would be well below the threshold of 10 million cubic meters annually specified in the Espoo Convention,[650] the Danube River Commission was notified of the project, and it provided a positive response.

The North-Western Sahara Aquifer System was revisited again, this time in a project in Tunisia, almost 20 years after the first project.[651] With the Consultation Mechanism fully established within OSS,[652] it was agreed that Tunisia could send the notification letter to the Mechanism for forwarding to both Algeria and Libya. As may be recalled, a similar approach is being followed with the NBI for Bank-financed projects on the Nile River Basin.

Transboundary groundwater was also considered in connection with the exceptions to the notification requirement. One water supply and sanitation project in Turkey included a component dealing with the rehabilitation of the existing water distribution networks in a number of towns.[653] The water supply source under the project came from groundwater wells and springs. The issues that arose were: (a) whether the wells and springs are connected to any international watercourses, and (b) if so, should those groundwater wells and springs be considered and dealt with as international waterways?[654] Based on the technical information provided, it was concluded that there was no connection between the groundwater wells and springs, and any international waterways. In addition to this finding, it was noted that even if surface water was involved, the proposed improvements would be

---

[650] *See supra* n. 366. This threshold, it may be recalled, requires also detailed environmental assessment, including assessment of the impact on the aquifer.

[651] Tunisia, Second Water Sector Investment Project (P095847, 2009). The Project included abstraction of water for irrigation in oases covering about 300 ha. It also included three deep boreholes to improve knowledge of the Aquifer, based on hydrogeological and geophysical studies. Although those wells would fall under the exception to the notification requirement regarding "water resources surveys and feasibility studies," it was decided that transparency and full disclosure require mentioning them in the notification letter, together with the other components of the project.

[652] *See supra* n. 637.

[653] *See* Turkey, Cesme-Alacati Water Supply and Sewerage Project (O08985, 1998).

[654] One of the questions raised initially in connection with this project was whether to adopt Barberis' definition of transboundary groundwater (*supra* n. 603) whereby a domestic aquifer receiving water from or contributing water to international surface water would qualify as transboundary groundwater.

exempt from the notification requirement under paragraph 7(a) of the policy because the project was only financing the rehabilitation of existing schemes. The project was finally processed under this exception.

In addition to those projects, the Bank carefully considers whether certain aquifers under some of its financed projects are shared with other states, or are wholly domestic or national aquifers.[655] The purpose of this exercise of due diligence is simply to ensure that the Bank has indeed acted prudently, in the interests of all the members of the institution, and to ascertain whether the project is governed by the practice of the Bank that has been established through the precedents discussed above.

Furthermore, the issue of notification was raised in the Guarani Aquifer Project,[656] but it was concluded that since the four riparian states of the Aquifer (Argentina, Brazil, Paraguay and Uruguay) would participate in, and are beneficiaries of the Project, and thus no notification under the Bank policy was actually needed.

Some of the concerns raised in connection with transboundary groundwater relate to the lack of reliable information about the number of international aquifers and the states sharing them, as well as the unavailability of technical information and data about those aquifers. However, two recent studies on transboundary groundwater should address some of those concerns. The first is the 2007 UNECE Assessment of Transboundary Rivers, Lakes and Groundwaters. The Assessment lists 140 transboundary rivers (most of them with a basin area over 1,000 cubic kilometers), and 30 transboundary lakes in the European and Asian parts of the UNECE region, as well as 70 transboundary aquifers in South-Eastern Europe, the Caucasus and Central Asia.[657] The second is the UNESCO

---

[655] In Ghana, Small Towns Water Supply and Sanitation Project (P084015, 2004) the Bank studied the data provided by the Ghana Water Resources Commission and agreed with the Commission that the aquifers to be utilized under the project are domestic ones. Similarly, in Angola, Third Social Action Fund (FAS III) (P081558, 2003), Hungary, Geothermal Energy Development Program (P075046, 2006), and Swaziland, Local Government Project (P095232, 2008) a study was commissioned in each case which concluded that the aquifer under each of the projects is domestic, and does not have any international connections.

[656] *See* Environmental Protection and Sustainable Development of the Guarani Aquifer System Project, *supra* n. 241.

[657] *See* Economic Commission for Europe, *Our Waters: Joining Hands Across Borders— First Assessment of Transboundary Rivers, Lakes and Groundwaters* (United Nations 2007). The Assessment has been undertaken by the parties and non-parties to the UNECE Convention on the Protection and Use of Transboundary Rivers and International Lakes (the Helsinki Convention). The UNECE has 56 countries, three of which are island states. The Assessment is available at: http://www.unece.org/env/water/publications/assessment/ assessmentweb_full.pdf.

International Hydrological Programme (IHP) Global Aquifer Map issued in 2008. The Map reveals that there are 273 transboundary aquifers: 68 in the Americas, 38 in Africa, 155 in Eastern and Western Europe and 12 in Asia.[658]

The discussion throughout this Chapter shows that transboundary groundwater is now well recognized as a shared resource. The overview of the Bank-financed projects indicates that the practice of notifying other riparians of projects that may affect transboundary groundwater is now clearly established, notwithstanding the fact that the definition of "international waterways" under the Bank policy does not include transboundary groundwater. The projects covered ranged from ones that depended exclusively on fossil aquifers, to others where the aquifers are connected to surface waters. Notification under those projects has been undertaken in some cases by the Bank and in others by the borrowers. The notified entities ranged from government ministries to river basin commissions. This process is similar in all accounts to the notification process for surface water discussed earlier. Yet, this process has been taking place through precedents and practice, rather than through an amendment to the policy to expand the definition of the term "international waterways" to include an explicit mention of transboundary groundwater. Such an amendment will simply codify the current practice of treating transboundary groundwater in the same manner as shared surface water for all the purposes specified under the Bank policy for projects on international waterways.

---

[658] The map also addresses the water quality of the aquifers, recharge of the aquifers, streams and rivers in the region of the aquifers, and population density near the aquifers. *See*: http://waterfortheages.wordpress.com/2008/10/26/unesco-launches-global-aquifer-map/.

CHAPTER 9

Linkages of the Policy to other World Bank Policies

9.1 Basis for the Linkages with other Policies

As the previous Chapters of this Book have explained, the Bank policy for projects on international waterways is based largely on two inter-related principles. The first is that the Bank must take into account the interests, not just of the borrower, but also of all its members. The basis for this principle is the requirement under the Bank's Articles of Agreement that the Bank should act prudently in the interests both of the particular member in whose territories the project is located and of the members as a whole.[659] The second principle, which is based on the first principle, is that the projects it is financing should not cause appreciable harm to other riparians.[660]

Application of these principles has not been limited to projects on international waterways where there are competing interests, demands and claims by the different riparians over the waters of the shared waterway. The Bank soon extended the same principles to projects in disputed areas. Such areas involve competing claims of sovereignty by different states, and include both land and waters. The rationale for extending the requirement of acting prudently from projects on international waterways to projects in disputed areas is the same. Just as two or more states may have competing claims over the shared waterway, two or more states may have conflicting claims over the sovereignty over the same area of land or waters. The disputed waters could be rivers, lakes, aquifers or maritime waters. The Bank as a financial cooperative development institution is under obligation to take into account those conflicting claims. This matter has been addressed in detail in a separate World Bank operational policy dedicated specifically to projects in disputed areas, as discussed below

The second principle requires that Bank-financed projects should not cause appreciable harm or adverse effects to any of the other riparians through water deprivation or pollution. As such the effects of the project could be quantitative

---

[659] *See supra* n. 84. As we have seen, OM 8 and the subsequent directives for projects on international waterways extended application of the policy to non-members as well. *See supra* n. 111.

[660] *See supra* Chapter 4, Part 4.5 of this Book.

or qualitative. However, the Bank policy for projects on international waterways does not define what would constitute appreciable harm or adverse effects. Furthermore, the policy does not define the term "pollution" or deal with environmental impact assessment. Those issues are addressed in details in a separate policy on the environment, discussed below.

The next two Parts of this Chapter will discuss and analyze the linkages between the Bank policy for projects on international waterways and those two policies, namely, projects in disputed areas, and environmental impact assessment.

9.2   Projects in Disputed Areas—Land and Water

As stated above, projects that could affect relations between the Bank and the borrower, as well as relations between the borrower and one or more neighboring countries, extend beyond those on international waterways. Projects in areas claimed by two or more states pose similar challenges to the Bank by virtue of the conflicting claims on sovereignty over such areas. The Bank recognized those challenges as early as 1983 when it issued its first directive addressing this matter—OMS 2.35, "Projects in Disputed Areas."

Indeed, that OMS equated the delicate problems that may arise as a result of financing a project in disputed areas to those arising out of projects on international waterways. The OMS stated in paragraph 1 that "'Projects in disputed areas may raise a number of delicate problems which (as in the case of projects on international waters—see Operational Manual Statement No. 2.32) could affect relations not only between the Bank and the borrower, but also between the borrower and one or more neighboring countries." It identified the areas that could be disputed and the problem of sovereignty thereon, stating that: "To the extent that projects involve physical investments or activities on land, on water, or in airspace, the Bank will hardly be able to avoid taking account of any issue which may arise relating to the sovereignty of the territories in which a proposed project is located."[661] Accordingly, the OMS clarified that disputed areas do not only cover land, but extend to water as well. Thus, disputed waters also fall under the ambit of disputed areas of the OMS.

It is important at the outset to distinguish between disputes over an international waterway, and disputes over sovereignty of the waterway itself or parts thereof. In the former case, the dispute, in most cases, is about the allocation of the waters of the shared river or lake, and how much water each of the riparians should get. When this kind of dispute intensifies and is brought to the forefront, it would affect the financing of a project on that waterway by the Bank.

---

[661] *See id.*, paragraph 2.

As discussed before, the Bank was not able to finance any projects on the Indus River System, either in India or Pakistan, until the dispute over the Indus River was resolved in 1960. Similarly, the dispute between Egypt and Sudan in the 1950s over the allocation of the Nile waters prevented the Bank from financing the Aswan High Dam Project in Egypt. The Roseiris Dam Project in Sudan was only financed after the 1959 Agreement was concluded between Egypt and the Sudan. As such, complex disputes would no doubt prevent the Bank from financing projects on the shared waterway. Other kinds of differences over the allocation of the waters of the shared waterway have not prevented the Bank from financing projects as long as the Bank can determine that the project would not cause appreciable harm to the other riparians.

It should be emphasized, however, that disputes over the sovereignty of the waterway itself or any parts thereof are far more complex and contentious, and go beyond the provisions of the Bank policy for projects on international waterways. Accordingly, those kinds of disputes are handled under the policy for projects in disputed areas.

An immediate question that arises is: what criteria is the Bank using to determine the existence of a dispute by two or more claimants over any area, whether land or water? GP 7.60 lays down four situations which would indicate the existence of such a dispute.

First, two or more countries have made public pronouncements claiming the same area. These pronouncements could be transmitted to the other country, third countries, regional organizations, or to the United Nations. Reference of the dispute to the United Nations and the issuance of a resolution or statement to that effect is one example.[662] Furthermore, adoption of statements or resolutions by the UN Security Council or the General Assembly would be a clear indication of the existence of such a dispute. Indeed, this is the most widely recognized method for ascertaining that a certain area is disputed. The dispute between India and Pakistan over Jammu and Kashmir was referred to the UN Security Council which issued a number of resolutions on the matter.[663]

Second, conflicting claims are made in the domestic legislation of two or more countries. Some countries define their boundaries in their constitutions or

---

[662] It should be clarified that the GP does not make any specific reference to resolutions of the UNGA or the Security Council. However, in the instances when the dispute is referred to the UNGA or the Security Council, a resolution or a statement would be issued under the UN Charter.

[663] Those resolutions are referred to in Security Council Resolution 91 (1951) Concerning the India-Pakistan Question, adopted by the Security Council on March 30, 1951 (Document No. S/2017/Rev. 1, dated March 30, 1951), available at: http://daccessdds.un. org/doc/RESOLUTION/GEN/NR0/072/10/IMG/NR007210.pdf?OpenElement.

legislation, and such demarcation may not comport with another country's demarcation of its boundaries. These kinds of conflicting claims may intensify, and may be taken to international or regional organizations for resolution. It could also remain dormant until it is ignited by another incident.[664]

Third, mineral, gas or oil exploration or similar rights are granted in the same area by two or more countries, or such rights granted by one country are protested by another. Such exploration rights or permits can be granted on land as well as on water. The dispute over the boundaries of the Caspian Sea within each of the five littoral states of Azerbaijan, Iran, Kazakhstan, Russia and Turkmenistan would fall under this category.[665] The dispute over the Caspian Sea among those states is exacerbated by the presence of considerable oil, gas and fisheries resources in the Sea, and the granting of exploration permits to a number of corporations by each littoral state.

Fourth, a dispute settlement procedure (e.g., arbitration, conciliation, mediation) is pending or differences continue after the completion of such procedure. The dispute between Ethiopia and Eritrea over their common borders was referred to the Permanent Court of Arbitration in December 2002, and the Eritrea-Ethiopia Boundary Commission delivered its decision on the delimitation of the borders of the two countries in April 2002.[666] However, differences on the

---

[664] The dispute between South Africa and Namibia with regard to their boundaries across the Orange River falls under this category. The Orange River forms the boundaries between the two countries for their entire borders across southern Namibia and northern South Africa. Namibia claims that the border should be in the middle of the Orange River. Indeed, Article 1(4) of the Namibian Constitution defines the national territory of Namibia as consisting of ". . . the whole of the territory recognized by the international community through the organs of the United Nations as Namibia, including the enclave, harbor and port of Walvis Bay, as well as the off-shore islands of Namibia, and its southern boundary shall extend to the middle of the Orange River." *See* Constitution of the Republic of Namibia, in *Constitutions of the Countries of the World* (Albert P. Blaustein & Gisbert H. Flanz, eds., Oceana Publications 2006). On the other hand, South Africa believes that the borders should be the deepest part of the river, which would be on the northern high-water mark, and not the middle of the river. In support of this claim, South Africa invokes and interprets the Anglo-German Treaty of 1890 as the basis for its claims. *See* Ian Brownlie, *African Boundaries—A Legal and Diplomatic Encyclopedia* 1273 (C. Hurst & Company 1979).

[665] Following the collapse of the Soviet Union in 1991, the new littoral states of the Caspian Sea—Azerbaijan, Russia, Kazakhstan, Turkmenistan and Iran—have been trying to negotiate a treaty related to the demarcation of the Caspian Sea among them. The demarcation would result in the agreement on the sharing, *inter alia*, of the considerable resources of oil, gas and fisheries in the Caspian Sea. However, no agreement has been reached yet on this matter, although the five states concluded the Framework Convention for the Protection of the Marine Environment of the Caspian Sea, November 14, 2003. *See supra* n. 252.

[666] The decision of the Eritrea-Ethiopia Boundary Commission is available at: http://www.pca-cpa.org/showpage.asp?pag_id=1150.

interpretation and implementation of the decision still persist, and large border areas between the two countries remain disputed.

Apart from the Caspian Sea and the Orange River, there are a number of examples of disputes over the sovereignty of shared waterways. The International Court of Justice (ICJ) has thus far adjudicated and decided three cases concerning border disputes across boundary rivers and lakes. Those cases involved, as mentioned earlier, Namibia and Botswana (the dispute over the Kasikili/Seddudu island on the Chobe River); Benin and Niger (Frontier dispute involving, *inter alia*, the Niger River); as well as Nigeria and Cameroon (involving, *inter alia*, their borders across the Lake Chad Basin).[667] A fourth dispute between Costa Rica and Nicaragua over the San Juan River is still pending before the ICJ.[668] As discussed earlier, the maritime limits of the Gulf of Fonseca were disputed by El Salvador, Honduras and Nicaragua, and the dispute over the Falklands/Malvinas between Argentina and the United Kingdom would by necessity involve the maritime boundaries and fishing rights in the area. Romania and Ukraine had a longstanding dispute regarding their maritime boundaries and the delimitation of the continental shelf and the exclusive economic zones in the Black Sea. The dispute was settled by the ICJ.[669] Other similar disputes include the Tanzania and Malawi dispute over the limits of Lake Malawi/Nyasa falling within each country.[670] Large stretches of the boundaries around the Mekong River between Lao PDR and Thailand, and between Cambodia and Thailand are still to be demarcated and agreed upon.[671] Any project proposed for Bank financing on any of those, or

---

[667] It should be clarified that boundaries, whether over land or water are, as a general rule, demarcated through treaties between the neighboring countries. Indeed, in each of three cases decided by the ICJ, there is a treaty demarcating the boundaries. The issue in the three cases is interpretation of the treaty. This also applies to the Costa Rica-Nicaragua dispute over the San Juan River. *See supra* n. 51.

[668] *See supra* n. 51.

[669] *See Maritime Delimitation in the Black Sea* (Romania v. Ukraine), available at: http://www.icj-cij.org/docket/index.php?p1=3&p2=3&code=ru&case=132&k=95.

[670] The Lake borders Malawi, Tanzania and Mozambique. *See* Brownlie, *supra* n. 664, at 1214.

[671] For more details on the border disputes between those countries *see* David Downing, *An Atlas of Territorial and Border Disputes*, 24 (New English Library 1980). It is worth adding in this connection that Cambodia and Thailand took their dispute over the Preah Vihear Temple which lies around the Mekong River to the ICJ in 1959, and the ICJ issued its decision on May 26, 1961 (*Temple of Preeh Vihear*, Cambodia v. Thailand). For more details on the dispute and the ICJ decision, *see*: http://www.icj-cij.org/docket/index.php?p1=3&p2=3&k=46&case=45&code=ct&p3=4. However, according to David Downing, the ICJ decision ". . . was not well received by Thailand, and there were serious clashes in the area in 1966." *See id.,* at 24. *See also* Caponera, *supra* n. 508, at 286.

similar waters would trigger both policies—OP 7.50, Projects on International Waterways, as well as OP 7.60 Projects in Disputed Areas.

OMS 2.35, issued in 1983, laid down detailed procedures for dealing with projects in disputed areas. That OMS was reissued in 1989 as OD 7.60 with some clarifications and elaborations. In 1994 the OD was converted into Operational Policy and Bank Procedures OP/BP 7.60, which were reissued in 2001 to reflect some organizational changes within the Bank.

The policies and procedures for projects in disputed areas elaborated in OP/BP 7.60 are based largely on the procedures for dealing with projects on international waterways, which in turn are predicated on the triangular relationship among the Bank, the borrower, and the other members of the Bank. Similar to projects on international waterways, the presence of any territorial dispute affecting a proposed Bank project is ascertained as early as possible and described in all project documents starting with the initial Project Information Document (PID). The policy requires that the territorial dispute be brought promptly to the attention of senior management of the Bank, who should be kept regularly informed of the processing steps of the project. To that effect a memorandum is prepared and sent to senior management which would (a) convey all pertinent information on the international aspects of the project, including information on the procedures followed and the outcome of any earlier projects the Bank may have considered in the disputed area; (b) make recommendations for dealing with the issue; and (c) seek approval for taking the actions recommended and for proceeding with project processing.[672]

On the substantive side, and also similar to the approach for projects on international waterways, the Project Appraisal Document (PAD) for a project in a disputed area discusses the nature of the dispute and affirms that Bank staff have considered it and are satisfied that either:

(a) the other claimants to the disputed area have no objection to the project;[673] or

(b) in all other instances, the special circumstances of the case warrant the Bank's support of the project notwithstanding any objection or lack of approval by the other claimants. Such special circumstances include the

---

[672] *See* paragraph 2 of BP 7.60.

[673] Paragraph 2 of OP 7.60 clarifies this by stating that "the Bank may support a project in a disputed area if the governments concerned agree that, pending the settlement of the dispute, the project proposed for country A should go forward without prejudice to the claims of country B."

[674] Two inter-related projects where such a determination was made were the Eritrea, Emergency Reconstruction Project (P044674, 2000) and the Ethiopia, Emergency

following (i) that the project is not harmful to the interests of other claimants,[674] or (ii) that the conflicting claim has not won international recognition or been actively pursued.[675]

Furthermore, the policy requires that in all cases, the project documentation bears a disclaimer stating that, by supporting the project, the Bank does not intend to make any judgment on the legal or other status of the territories concerned, or to prejudice the final determination of the parties' claims.[676]

A number of observations can be made on this paragraph:

First, similar to the approach for projects on international waterways, the Bank can proceed with the processing and financing of the project if the other claimants to the disputed area have no objection to the project. For these claimants to voice their no-objection, they need to be notified of the project, provided with the project details, and given time to respond, just as the policy for projects on international waterways prescribes. However, the policy for projects in disputed areas does not include any procedures for notification. Indeed, the term "notification" is not even mentioned in the policy. The question that immediately arises is how, in the absence of provisions on notification, has the Bank been able to ascertain that the other claimants have no objection to the project? As may be recalled, this was the same situation faced in connection with OM 5.05 of 1965.[677]

It must be emphasized that projects in disputed areas pose far more challenges than projects on international waterways. Most of the countries involved in such disputes would not even admit that the area in question is really disputed, and few

---

Reconstruction Projects (P067084, 2000). Both countries requested the Bank to assist in the reconstruction and rehabilitation of the areas affected by the war that lasted for two years between them (1998–2000). Those areas were the disputed areas that caused the war in the first place, and thus OP 7.60 was applicable to both projects. Neither party was willing to give its no-objection to the project of the other party, but neither objected to the other project. The Bank decided to proceed with the processing of the projects under paragraph 3(a)(i)—that neither project was harmful to the interests of the other party. The basis of this determination was that both countries would receive similar treatment, and that the proceeds of the IDA Credits for both countries would be used to finance the rehabilitation of critical socio-economic infrastructure destroyed by the war, and the provision of essential agricultural inputs to resume agricultural production. This justification is similar to the exception to the notification requirement dealing with the rehabilitation of existing schemes under paragraph 7(a) of OP 7.50.

[675] *See* paragraph 3 of OP 7.60.

[676] *See id.*

[677] *See supra*, Chapter 2, Part 2.3 of this Book. Neither OM 8, nor OM 5.05 included a mention of the term "notification." OM 5.05, however, included provisions on the no-objection of other riparians, as discussed earlier.

would agree to refer the dispute to a third party for resolution.[678] The claim by both parties, in most cases, is that the area in question (whether it is land or water) is theirs, and that the claim of the other side is without basis. That makes the starting point for addressing the issue quite complex. This is perhaps one reason why detailed procedures on the notification of the other claimants to the disputed areas have not been spelled out in the policy, leaving the issue to be addressed on a case-by-case basis.

On the other hand, most of the riparians would acknowledge that the other riparians have rights over the shared waterway. The point of contention in those kinds of cases is: which riparian gets how much of the waters of the shared river or lake, and when do the effects of the project amount to appreciable harm. In this situation, there is in most cases, a considerable space for the Bank to move in and finance the project without causing harm to other riparians.

Similar to projects on international waterways, the Bank has been able since the first policy for projects in disputed areas was issued in 1983, to develop detailed procedures for dealing with projects in disputed areas, and to fill the gaps in its directives. Those procedures have been gradually refined and streamlined through practice. The procedures consist of discussing the project in the disputed area, at a very early stage, with the Executive Directors concerned, and thus to gauge their position *vis-à-vis* the project.[679] If no objection is voiced, the Bank would proceed with processing the project. If an objection is conveyed to the Bank, then the Bank, at least theoretically, can still support the project if the special circumstances of the case warrant such support. Those special circumstances are either that (i) the project is not harmful to the interests of other claimants, or (ii) the conflicting claim has not won international recognition or been actively pursued by the other party.

It is noteworthy that the policy uses the term "harmful" which is in line with the term "appreciable harm" used in connection with projects on international waterways. The policy allows the Bank to make a determination that the project

---

[678] The Guatemala-Belize dispute can be considered as one of the exceptions to this general rule. Guatemala has for sometime been claiming large territories of Belize. On December 8, 2008, the two countries signed an agreement which, subject to approval in a referendum in each country, would refer the dispute to the ICJ. For a copy of the Agreement, *see:* http://www.oas.org/sap/docs/fondo_paz/AR-M450U_20081208_145042.pdf. For an explanatory note of the Agreement *see:* http://www.governmentofbelize.gov.bz/press_release_details.php?pr_id=5235.

[679] The Argentina, Sustainable Fisheries Management Project (P057459, 2000) has been one of the projects where these procedures were followed. The Project aimed at strengthening the monitoring and control of commercial fishing within the entire exclusive economic zone (EEZ) claimed by Argentina. This EEZ includes the Falklands/Malvinas Islands which are claimed by both Argentina and the United Kingdom, and accordingly OP 7.60 applied to the Project. The Executive Director representing the United Kingdom

"is not harmful to the interest of other claimants," and this has provided a space for making a judgment. One of the projects where the Bank exercised that judgment is the Azerbaijan Petroleum Technical Assistance Project.[680] The proposed technical assistance would finance advisory services which involved the rehabilitation and upgrading of existing petroleum facilities of Azerbaijan in the Caspian Sea. As indicated above, five states emerged as the littoral states of the Caspian Sea, following the collapse of the Soviet Union in 1991. Those states are the Russian Federation, Iran, Azerbaijan, Kazakhstan and Turkmenistan. No agreement has been reached on the demarcation of the Caspian Sea among those states, and there have been varying claims by each of them. Consequently the Bank has concluded that the Caspian Sea is a disputed area under the Bank OP 7.60. The Bank has also concluded that the Caspian Sea is an international waterway under OP 7.50.[681] Thus, the two policies were invoked for this project. This is perhaps the first time, and one of the few instances, where the two policies were applied for the same project.

However, because of the nature of the activities financed under the project which consisted primarily of studies and the rehabilitation of existing oil facilities, the Bank concluded that the project would have no adverse effects on any of the littoral states, and thus would fall under the exception to the notification requirement under paragraph 7(a) of the policy for projects on international waters (rehabilitation of existing schemes). That being said, the same conclusion was arrived at with regard to the policy for projects in disputed areas, namely that the project was not harmful to the interests of the other riparians.[682] Nonetheless, the Project documents included the disclaimer that, by supporting the project, the Bank does not intend to make any judgment on the legal or other status of the territories concerned or to prejudice the final determination of the parties' claims.[683] The issues under this project were, however, made easier by the fact that the project fell under the exception to the notification requirement.

The other situation which would allow the Bank to finance a project in a disputed area is that the conflicting claim has not won international recognition or been actively pursued. The claims of Somalia to territories which are currently

---

was consulted about the Project, and following bilateral discussions between the two countries, the Executive Director gave a no-objection to the Project.

[680] Azerbaijan, Petroleum Technical Assistance Project (P008282, 1995).

[681] *See supra* n. 294.

[682] In making this determination, the Bank took note of the fact that Azerbaijan has been disposing oil from the oil field under the Project, as well as from other offshore fields for its own account since gaining independence in 1991, without any objection from the other Caspian Sea states.

[683] *See* paragraph 3 of OP 7.60.

part of Ethiopia and Kenya[684] would likely fall under this category since the international community has not recognized those claims, nor has Somalia itself actively pursued any of them recently.

The second observation relates to the use of the term "claimants" in paragraph 3 of the policy. The use of this term would allow the policy to cover situations where the dispute is not between two governments, but between one government and a nationalist movement, as happened in East Timor before it became independent, or SWAPO and South Africa over Namibia.[685] In such cases, the dispute is actually between two claimants, and one of them is not a government. As may be recalled, the Bank asked Lesotho to consult with the SWAPO on the Lesotho Highlands Water Project under OP 7.50.[686] This request was made notwithstanding the use by the policy of the terms "neighboring countries" and "governments concerned"[687] in another paragraph, which is quite appropriate when the dispute over the area in question is between two neighboring states.

The third observation concerns the requirement under the policy of inclusion of the disclaimer that by supporting the project, the Bank does not intend to make any judgment on the legal or other status of the territories concerned or to prejudice the final determination of the parties' claims. Inclusion of this disclaimer has not always been an easy task. As stated earlier, most of the claimant states would not consider the areas in question as disputed, and because of that even this neutral language is opposed by many of them.

One other area addressed by the two policies is the use of maps. Inclusion of maps would usually pose some difficulties because no state likes to have part of its territory shown as disputed, even when this fact is an established one. Unlike the policy for projects on international waterways, the policy for projects in disputed areas does not make an explicit requirement for inclusion of maps in any project documents. The requirement for inclusion of maps in these types of projects stems from the general approach of the Bank to include in the PAD a map of the project location. However, both policies require that maps be omitted after management permission is obtained.[688]

Clearly, projects in disputed areas raise some complex issues, and could subject the Bank to challenges from the countries concerned. Those challenges are more difficult and complex when the project falls on an international waterway over which sovereignty and ownership is claimed by two or more states.

---

[684] *See* Downing, *supra* n. 671, at 66–70.

[685] OP 7.60 also uses the term "parties" in the disclaimer in paragraph 3(b)(ii) where it refers to the "parties' claims."

[686] *See supra* n. 429.

[687] *See* paragraph 2 of OP 7.60.

[688] *See* paragraph 13 of OP 7.50, and paragraph 6 of OP 7.60.

9.3 Environmental Aspects of Projects on International Waterways

The absence of any provisions in the policy for projects on international waterways on the environment generally, or on environmental impact assessments in particular, is worth noting. None of the directives on projects on international waterways issued since 1956 have provided any details on how to determine if the project has the potential of causing harm through water deprivation or pollution; nor have they defined the term "pollution" or explained what it entails.[689] These directives have also not clarified the situations or manner in which water quality and quantity could be adversely affected. Such matters have been left to be determined by principles, guidelines and standards outside the scope of the policy for projects on international waterways.

As discussed earlier, the first Bank policy on the environment, OMS 2.36, "Environmental Aspects of Bank Work" was issued in 1984, almost 30 years after the first Bank policy for projects on international waterways was adopted.[690] The absence of a policy on the environment until that time was perhaps one reason as to why semi-enclosed waters were added in 1965 to the list of waters covered under the policy for projects on international waterways.

The 1984 OMS on the environment laid down very basic and rudimentary requirements. It instructed the Bank staff to identify as early as possible projects that would have significant environmental effects, and determine the investigations and measures needed to prevent or mitigate any serious adverse effects of each such project. That OMS and the subsequent directives, including Annexes to OD 4.01 and other memoranda on environmental assessment, were all replaced in 1999 with OP/BP 4.01 on Environmental Assessment. The OP/BP includes a number of annexes,[691] and has been complemented by other guidebooks and sourcebooks in the field of environment.[692]

---

[689] For definition of the term "pollution" under different instruments, *see infra* n. 699.

[690] *See supra* n. 108. It is worth noting that the first policy on the environment, OMS 2.36, was issued one year before OMS 2.32, Projects on International Waterways, was issued in 1985.

[691] Annex A to the OP deals with definitions, Annex B deals with the content of an Environmental Assessment Report for a Category A project, and Annex C with Environmental Management Plans. Application of the EA to dam and reservoir projects is addressed in Annex B to the BP, while application of the EA to projects involving pest management is dealt with in Annex C to the BP.

[692] Two leading publications in this field are the *Pollution Prevention and Abatement Handbook* (The World Bank 1999), and the *Environmental Assessment Sourcebook* (The World Bank 1999). As indicated earlier, those guidebooks/sourcebooks are gradually replacing the Good Practices that were issued in 1994, complementing the OPs and BPs.

OP 4.01 states that the Bank requires preparation of an environmental assessment (EA) of projects proposed for Bank financing to help ensure that they are environmentally sound and sustainable, and to address any potential adverse impacts.[693] It defines the EA as a process whose breadth, depth, and type of analysis depend on the nature, scale, and potential environmental impact of the proposed project.[694] The EA evaluates a project's potential environmental risks and impacts in its area of influence, examines project alternatives;[695] identifies ways of improving project selection, siting, planning, design, and implementation by preventing, minimizing, mitigating, or compensating for adverse environmental impacts. The OP requires the EA to take into account the natural environment (which it defines to include air, water, and land); human health

---

[693] The Bank classifies the proposed project into one of four categories, depending on the type, location, sensitivity, and scale of the project and the nature and magnitude of its potential environmental impacts:

(a) *Category A*: A proposed project is classified as Category A if it is likely to have significant adverse environmental impacts that are sensitive, diverse, or unprecedented. These impacts may affect an area broader than the sites or facilities subject to physical works.

(b) *Category B*: A proposed project is classified as Category B if it's potential adverse environmental impacts on human populations or environmentally important areas—including wetlands, forests, grasslands, and other natural habitats-are less adverse than those of Category A projects. These impacts are site-specific; few if any of them are irreversible; and in most cases mitigatory measures can be designed more readily than for Category A projects.

(c) *Category C*: A proposed project is classified as Category C if it is likely to have minimal or no adverse environmental impacts. Beyond screening, no further EA action is required for a Category C project.

(d) *Category FI*: A proposed project is classified as Category FI (Financial Intermediary) if it involves investment of Bank funds through a financial intermediary, in sub-projects that may result in adverse environmental impacts. Through practice, classification of projects as FI has been extended to all investments that are not known at the time of appraisal, and which would be selected during implementation, whether carried out through financial intermediary or another entity.

[694] Annex A to the OP defines an Environmental Impact Assessment (EIA) as "An instrument to identify and assess the potential environmental impacts of a proposed project, evaluate alternatives, and design appropriate mitigation, management, and monitoring measures. Projects and sub-projects need EIA to address important issues not covered by any applicable regional or sectoral EA."

[695] The analysis of alternatives compares feasible alternatives to the proposed project site, technology, design, and operation—including the "without project" situation—in terms of their potential environmental impacts; the feasibility of mitigating these impacts; their capital and recurrent costs; their suitability under local conditions; and their institutional, training, and monitoring requirements. *See* Annex B of OP 4.01. *See* also *supra* n. 646.

and safety; social aspects (involuntary resettlement, indigenous peoples, and physical cultural resources), as well as transboundary and global environmental aspects. The OP clarifies that such global environmental issues include climate change, ozone-depleting substances, pollution of international waters and adverse impacts on biodiversity.[696]

Annex A to the OP defines both, the Regional EA, as well as the Sectoral EA. The former is an instrument that examines environmental issues and impacts associated with a particular strategy, policy, plan or program, or with series of projects for a particular region (such as an urban area, a watershed or a coastal zone).[697] The Annex also requires that the Sectoral EA examines environmental issues and impacts associated with a particular strategy, policy, plan or program or with a series of projects for a specific sector (such as power, transport or agriculture). The Annex to the OP requires that both the Regional EA and the Sectoral EA pay particular attention to potential cumulative impacts of multiple activities.

Thus, OP 4.01 clearly highlights international waters, and requires that a range of environmental issues relating to projects on international waterways be comprehensively addressed. Those issues go beyond the basic and rudimentary requirement under the policy for projects on international waterways of determining if the project has the potential of causing harm to other riparians through water deprivation or pollution.[698] In addition, the OP provides detailed definition and guidance on those issues, and requires evaluating and comparing the impacts against those of alternative options.

Indeed, the notification letters and Project Details sent to other riparians now routinely refer to the EA, highlight its main findings, and use those findings to draw the conclusion that the project would not cause appreciable harm to any of the riparians. The *Environmental Assessment Sourcebook* underscores clearly the

---

[696] *See* OP 4.01, n. 4.

[697] *See id.*, Annex A.

[698] The World Bank Group *Pollution Prevention and Abatement Handbook* defines the term "Pollution" to mean "Generally, the presence of matter or energy whose nature, location, or quantity produces undesired environmental effects." *See Pollution Prevention and Abatement Handbook*, *supra* n. 692, at 452. The Handbook goes on to state that "Under the US Clean Water Act, for example, the term is defined as the man-made or man induced alteration of the physical, biological, and radiological integrity of water." The UN Watercourses Convention follows the approach of "man-made" and defines "pollution of an international watercourse" to mean "any detrimental alteration in the composition or quality of the waters of an international watercourse which results directly or indirectly from human conduct." *See* Article 21(1) of the Convention. Article 1 of the UNCLOS follows a similar approach to the definition of the term "pollution," *see supra* n. 286. *See* also the IIL definition of the term in *supra* n. 160.

integration of the environmental assessment process and notification when it states that:

> The environmental impacts of projects always should be evaluated as early as possible in planning, but it is crucial when international waterways are concerned. When the other riparians are notified of a project that may involve their waterways, adequate information and data should be provided to enable them to determine the potential effects of such a project.[699]

In the instances where one of the riparians has requested additional information on the project, such information would usually relate to the EA. In fact, in many instances, copies of the EA itself have been requested by the notified riparians. A point was made by some of those riparians that they would not be in a position to make a determination about the effects of the project without receiving and studying the EA. Other questions raised by the notified riparians typically refer to EA issues, such as cumulative impacts of the previous projects on the same waterway, as well as impacts of the project in question on fisheries and the aquatic systems.

The requests for the full EA have raised issues with regard to national projects and investments where the sub-projects to be funded under the Bank-financed projects would be known only during implementation. Under those types of projects, the full EA cannot be prepared by the time of appraisal of the project because those sub-projects would be selected during implementation. Consequently, the Bank requires preparation of an environmental policy framework which will be used to prepare a full EA for each sub-project. Concerns have been raised by some of the notified riparians that the information contained in the policy framework alone would not be adequate to enable them to determine if the project would cause any harm, nor would it help them ascertain the extent of the harm that the project may cause. On the other hand, the policy requires that the process of notification be completed by the time the project is presented to the Executive Directors. The Bank and the borrowers have tried to address this situation by providing additional information related to the environmental aspects of the project, such as environmental due diligence and other environmental studies on the project.

However, not all the elements of the policy on environmental assessment can be extended to projects on international waterways. The provisions of the environmental assessment policy on disclosure and public consultations are examples of provisions that cannot have extra-territorial application. Disclosure[700] and public

---

[699] *See* World Bank, *Environmental Assessment Sourcebook*, 1999, Chapter 2; available at: http://go.worldbank.org/D10M0X2V10.

[700] With regard to disclosure, paragraph 15 of OP 4.01 states that "For meaningful consultations between the borrower and project-affected groups and local NGOs on all Category

consultations[701] in connection with a project on an international waterway can only be undertaken by the borrower within its own state and cannot be carried in another country, even if that country is a riparian. Accordingly, project-affected groups and local non-governmental organizations outside the borrower's country cannot be consulted on the effects of a project on international waterways on such groups by the borrower. Such groups and organization may have access to some information about the project through the information disclosed globally by the Bank.[702] They could also have information on the project through the notification letter and Project Details provided to their government, if the government wishes to disclose such information and consults with them.

Hence, the close linkages between the two policies are quite clear, and indeed necessary. Although OP 7.50 does not make any reference to OP 4.01 or to environmental assessment, notification letters and Project Details now routinely make such a reference. The EIA analysis and findings are used to confirm the Bank's conclusion that the project will not cause any appreciable harm to the other riparians. As indicated earlier, almost all of the requests for additional information have dealt with environmental issues, and sometimes specifically asked for a copy of the EIA.

Those linkages became quite clear in the different reports issued in connection with the request for inspection of the Bujagali Project in Uganda[703] submitted to the World Bank Inspection Panel.[704] The Project consists of the construction of the

---

A and B projects proposed for IBRD or IDA financing, the borrower provides relevant material in a timely manner prior to consultation and in a form and language that are understandable and accessible to the groups being consulted."

[701] With regard to public consultations, paragraph 14 of OP 4.01 states that "For all Category A and B projects proposed for IBRD or IDA financing, during the EA process, the borrower consults project-affected groups and local nongovernmental organizations (NGOs) about the project's environmental aspects and takes their views into account. The borrower initiates such consultations as early as possible. For Category A projects, the borrower consults these groups at least twice: (a) shortly after environmental screening and before the terms of reference for the EA are finalized; and (b) once a draft EA report is prepared. In addition, the borrower consults with such groups throughout project implementation as necessary to address EA-related issues that affect them." For further discussion and analysis of the matter, *see* Charles E. Di Leva, *Access to Information, Public Participation and Conflict Resolution at the World Bank*, in *Public Participation in the Governance of International Freshwater Resources*, 199–215 (Carl Bruch *et al.*, eds., United Nations University Press 2005).

[702] Paragraph 18 of OP 4.01 states that once the borrower officially transmits the EA report to the Bank, the Bank distributes the summary (in English) to the Executive Directors (EDs) and makes the report available through the Bank's InfoShop.

[703] Uganda, Private Power Generation (Bujagali) Project (P089659, 2008).

[704] The Inspection Panel was established through IBRD Resolution No. 93–10, and the identical IDA Resolution No. 93–6, both adopted by the Executive Directors of the respective

Bujagali run-of-river hydropower plant on the Nile River near the Bujagali Falls,[705] about eight kilometers downstream from the existing Kiira-Nalubaale Hydropower Plants, where the Nile exits Lake Victoria.[706] It is designed for a generation capacity of about 250 megawatts (MW) of power. The Project is a public-private partnership between private sponsors and the Government of Uganda. It is supported by private lenders and multilateral and bilateral development agencies, including the World Bank Group, the European Investment Bank, and the African Development Bank. The total cost of the project is estimated to be around US$860 million. The World Bank Group support is about US$360 million, including a partial risk guarantee from IDA of US$115 million, loans from IFC, and a guarantee from MIGA.[707] The IDA Guarantee was approved by the Executive Directors on April 26, 2007.

---

institution on September 22, 1993. The purpose of the Panel is to serve as an independent mechanism for ensuring accountability in Bank operations through assessing compliance by the Bank with its operational polices and procedures. A group of two or more private citizens who believe that they or their interests have been or could be harmed by Bank-financed projects can present their concerns to the Panel. As of January 2009, 52 formal requests have been received since Panel operations began in September 1994. For more details on the Inspection Panel, *see*: http://www.inspectionpanel.org. *See* also Ibrahim Shihata, *The World Bank Inspection Panel* (2d ed., Oxford 2001); and David Freestone, *The Environmental and Social Safeguard Policies of the World Bank and the Evolving Role of the Inspection Panel*, in *Economic Globalization and Compliance with International Environmental Agreements*, at 139 (Kanami Ishibashi, Alexandre Kiss & Dinah Shelton, eds., Kluwer Law International 2003).

[705] The Bujagali Project consists of a run-of-river hydropower power plant with a reservoir for daily storage, an intake powerhouse complex, and an earth filled dam with a maximum height of about 30 meters, together with spillway and other associated works. The reservoir will have an estimated surface area of about 388 hectares, extending back to the tailrace areas of the Nalubaale and Kiira dam complex. The Project also includes the construction of about 100 kilometers of transmission lines.

[706] Lake Victoria is the largest freshwater body in the developing world and the second largest lake in the world after Lake Superior in North America. It covers a surface area of about 68,870 square kilometers. About 49 percent of the surface area of Lake Victoria falls within the boundaries of Tanzania, 45 percent within Uganda, and the remaining 6 percent is within Kenya. The Lake is the source of about 15 percent of the waters of the Nile River; with the remaining balance coming from the Ethiopian plateau and Lake Tana. *See* Lake Victoria Development Programme, available at: http://www.eac.int/LVDP/basin.htm.

[707] This is the second attempt by the Bank to finance the Bujagali Project. The first project terminated in September 2003 because of the withdrawal of the private sponsor. That project was also the subject of an Inspection Panel request in July 2001, together with two other Bank-financed projects in Uganda, Third Power Project (also referred to as Owen Falls Extension, now known as Kiira, and the Owen Falls known as Nalubaale), and Fourth Power Project. Bank Management responded to that Request in September 2001, but the Executive Directors approved the Panel's recommendation for investigation in October 2001. The Panel issued its Investigation Report on that request in May 23, 2002, and the Management Report and Recommendation was issued in June 2002. The Executive Directors approved the Report and Recommendation in June 2002.

The request for inspection of the Project was received on March 5, 2007, about two months before the project was approved by the Executive Directors.[708] The request contended that the Bank has failed to follow a number of its operational policies and procedures in the design and appraisal of the Project,[709] and that this failure would result in serious harm to the people living in the Project area and to the environment, in particular the Nile River and Lake Victoria,[710] and to the customers of the generated electricity, and to Uganda citizens in general. Bank Management issued its Response in April 2007, addressing the issues raised by the request for inspection. However, the Executive Directors approved the Inspection Panel's recommendation to conduct an investigation in May 2007. The Panel issued the Investigation Report in September 2008, and Bank management issued its Report and Recommendation on November 7, 2008.[711] On December 4, 2008, the Executive Directors discussed both documents and approved the Management Report and Recommendation.[712]

Those reports are available at: http://web.worldbank.org/WBSITE/EXTERNAL/EXTINSPECTIONPANEL/0,,contentMDK:20228304~pagePK:64129751~piPK:64128378~theSitePK:380794,00.html.

[708] A similar request was submitted to the Compliance Review and Mediation Unit (CRMU) of the African Development Bank (AfDB) that is also co-financing the Bujagali Project. For the details of the request and the action taken thereon, *see*: http://www.afdb.org/portal/page?_pageid=473,19836271&_dad=portal&_schema=PORTAL. For a detailed discussion and analysis of the accountability mechanisms of the different international financial institutions, including the Inspection Panel and the CRMU, *see* Maartje Van Putten, *Policing the Banks—Accountability Mechanisms for the Financial Sector* (McGill-Queen's University Press 2008).

[709] Paragraph 12 of the Resolution establishing the Inspection Panel states that "The affected party must demonstrate that its rights or interests have been or are likely to be directly affected by an action or omission of the Bank as a result of a failure of the Bank to follow its operational policies and procedures with respect to the design, appraisal and/or implementation of a project financed by the Bank. . . ." Because design and appraisal of the project fall under the mandate of the Panel, the Request for Inspection can be submitted before the Executive Directors approve the project, as happened in this case.

[710] The requesters claimed that the water level of Lake Victoria has declined in recent years because of the drought and withdrawal for power under Kiira and Nalubaale projects, and that there would not be enough water for the Bujagali project.

[711] Following registration of any request by the Inspection Panel, a copy is sent to the Bank Management and the Executive Directors. Bank Management prepares the "Management Responses" after which the Panel determines whether to request the Executive Directors to authorize an investigation of the project. If an investigation is authorized, the Panel undertakes such investigation and prepares the "Investigation Report." Management responds by preparing the "Management Report and Recommendation" which includes an action plan for addressing the issues raised by the request and the Panel. Both documents are discussed by the Executive Directors, and disclosed thereafter.

[712] *See* Inspection Panel Investigation Report (Report No. 44977-UG, dated August 29, 2008), and Management Report and Recommendation in Response to the Inspection

Based on the request for inspection, the Panel assessed whether the Bank has complied with 12 of its policies, including OP/BP 4.01 on Environmental Assessment, OP/BP 7.50, Projects on International Waterways, and OP/BP 4.07 on Water Resources Management.[713] The Panel examined, among many issues, a range of hydrological and environmental issues, in connection with the request for inspection and the above cited three policies, including the impact of hydrologic risk on energy output, the potential impact of the Project on the levels of Lake Victoria, and the risks of climate change.[714]

The Panel noted the substantial body of analysis under the Project, and found that the hydrologic data sets used in the Project design constitute a reliable data series and an appropriate baseline for analysis in compliance with OP 4.01. The Panel Investigation Report, however, stated that there were a number of areas of non-compliance with OP 4.01 in relation to Lake Victoria and the Nile. Those areas, according to the Investigation Report, included:

First, the Project Social and Environmental Assessment (SEA) did not adequately make reference to the Strategic/Sectoral Social and Environmental Assessment (SSEA) of the separate Nile Basin Initiative (NBI), which analyzed issues such as climate change and cumulative effects.[715] As a result, important information required under Bank policy was not disclosed in a timely manner as an integral part of the Project's documentation.

Panel Investigation Report (Report No. IDA/R2008-0296). These Reports, together with the Request for Inspection, Management Response, and the Eligibility Report, are available at: http://web.worldbank.org/WBSITE/EXTERNAL/EXTINSPECTIONPANEL/0,,contentMDK:21247695~pagePK:64129751~piPK:64128378~theSitePK:380794,00.html.

[713] The 12 policies whose compliance was assessed by the Panel were: OP 1.00 Poverty Reduction; OP/BP 4.01 Environmental Assessment; OP/BP 4.02 Environmental Action Plans; OP/BP 4.04 Natural Habitats; OP 4.07 Water Resource Management; OP/BP 4.10 Indigenous Peoples; OP/BP 4.11 Physical Cultural Resources; OP/BP 4.12 Involuntary Resettlement; OP/BP 4.37 Safety of Dams; OP/BP 7.50 Projects on International Waterways; and OP/BP 10.04 Economic Evaluation of Investment Operations, in addition to the World Bank Policy on Disclosure of Information.

[714] Discussion and analysis of those issues are beyond the scope of this Part of the Book. This Part merely intends to clarify the linkages between OP 4.01 and OP 7.50.

[715] The Panel referred to the World Bank–Netherlands Water Partnership (BNWPP), background description for the "Victoria Nile-Independent Hydrological Review" activity, available at: http://www-esd.worldbank.org/bnwpp/index.cfm?display=display_activity&AID=439. Established in 2000 by the Government of the Netherlands and the World Bank, the BNWPP mission is to improve water security by promoting innovative approaches to Integrated Water Resources Management (IWRM), and assist developing countries in managing their water resources in ways that promote environmentally sustainable growth, poverty reduction and gender equity. For more information on the BNWPP, *see*: http://www-esd.worldbank.org/bnwpp/.

Second, neither the SSEA nor the SEA addressed the cumulative effects of the existing and planned projects in a meaningful way. The SEA also considered that the Project's area of influence ends downstream of the Kiira-Nalubaale dams and therefore did not assess the Project's potential impacts on the changing levels of Lake Victoria, as it should have. This is particularly important because the lowering of water levels in the Lake, as has occurred in recent years, causes significant social and environmental impacts.

Third, the PAD's categorical assertion, without any reference to risk and uncertainty, that there would be no adverse effect on water release due to climate change during the Project life failed to express a potential risk factor, which was identified in the SSEA, as required.

Although the Panel stated that it assessed, *inter alia*, compliance with OP 7.50, it did not make any reference to compliance or non-compliance with this OP, nor with OP 4.07 on Water Resources Management. Indeed, Uganda itself notified the other nine Nile riparian states of the initial project in 2000. In March 2007, new notification letters, updating the information contained in the earlier letters and the Project Details, were sent again to the nine riparian states as well as to the NBI Secretariat.[716] The notification letters confirmed the Government of Uganda's intention to continue with the present operating procedure which has been used since 1954,[717] known as the "Agreed Curve," to determine and produce the controlled discharge down the Victoria Nile River.[718] Consequently, the Government noted that the Bujagali hydropower plant would not result in any change

---

[716] Unlike other recent Bank-financed projects that affect the Nile, the NBI was not the mechanism that sent the notification letters to the riparian states. Rather, the NBI was informed of the project and of the notification to the other riparians. Thus, no action was expected from the NBI.

[717] Construction of the Owen Falls Dam (now known as the Nalubaale Dam) started in 1949, based on three agreements: (i) Exchange of Notes Constituting an Agreement between the Government of the United Kingdom of Great Britain and Northern Ireland and the Government of Egypt Regarding the Construction of the Owen Falls Dam, Cairo, May 30–31, 1949, 226 U.N.T.S. 274 (1956); (ii) Exchange of Notes Constituting an Agreement between the Government of the United Kingdom of Great Britain and Northern Ireland and the Government of Egypt Regarding the Construction of the Owen Falls Dam, Cairo, December 5, 1949, 226 U.N.T.S. 280 (1956); and (iii) Exchange of Notes (with enclosure) Constituting an Agreement between the Governments of the United Kingdom of Great Britain and Northern Ireland and the Government of Egypt Regarding the Construction of the Owen Falls Dam, Cairo, July 16, 1952, and January 5, 1953, 207 U.N.T.S. 278 (1955). For more details on the dam and those agreements, *see* C. O. Okidi, *History of the Nile and Lake Victoria Basins Through Treaties* in *The Nile—Resources Evaluation, Resource Management, Hydropolitics and Legal Issues*, 203 (P. P. Howell & J. A. Allan, eds., School of African and Oriental Studies, University of London 1990).

[718] Before completion of the Owen Falls Dam in 1959, the outflow of the Victoria Nile from Lake Victoria was naturally regulated by the Rippon Falls. However, completion of

to the discharge pattern in the Victoria Nile River with respect to the range of water levels of Lake Victoria that would have occurred had this hydropower facility not been constructed.[719] There was no unfavorable response to the notification, apparently because the project was discussed at one of the Nile Ministerial Council meetings, where no concerns were voiced.

This determination was confirmed by both, the Management Response, as well as the Management Report and Recommendation (the Report). The Report stated that since the Bujagali Hydropower Plant is downstream, it will reuse the water coming from the upstream complex. Specifically, the water passing through Nalubaale and Kiira and subsequently through Bujagali will produce more than twice the amount of energy that Nalubaale and Kiira would produce alone, and this is expected to lead to a more efficient use of water for power generation in line with the Agreed Curve. Thus, during droughts, Bujagali would enhance Lake management as well as power production.

The Report further stated that water flow arriving at the Bujagali Hydropower facility has been, and will continue to be, completely controlled by discharges from the mouth of the Nile River at the Nalubaale/Kiira hydropower dam complex, which is eight kilometers upstream of the Bujagali Dam site.[720] Hence, the operators of the Bujagali facility will have no control over water releases from

---

the dam resulted in removing this natural barrier, and converting the Lake into a reservoir. Outflow from the Lake is now regulated by the Agreed Curve, which was arrived at between Egypt and the British colonial administration in Uganda, as a result of the agreements referred to above (*supra* n. 717). The Agreed Curve is a mathematical formula that correlates the flow of the Nile at the source from Lake Victoria to the water level in the Lake, basically to mimic the natural flow. For more discussion of the topic *see* J. V. Sutcliffe & Y. P Parks, *The Hydrology of the Nile*, IAHS Special Publication No. 5 (International Association of Hydrological Sciences 1999) at 28. The sources of the waters of the Lake consist of river flows in the Lake (the Kagera River being the most important such source), groundwater inflow and rainfall, balanced by evaporation and the Victoria Nile outflow.

[719] It should be added that the three states bordering Lake Victoria, Kenya, Tanzania and Uganda concluded the Protocol for Sustainable Development of Lake Victoria Basin on November 29, 2003. The Protocol is a comprehensive instrument under which the three parties agreed to cooperate in a number of areas. Such areas include the conservation and sustainable utilization of the resource of the Lake, including the sustainable development, management and equitable utilization of water resources, fisheries, agriculture and land use practices, forests, management of wetlands; trade, commerce and industrial development; infrastructure and energy; navigational safety and marine security; and promotion of research, capacity building and information exchange. The Protocol entered into force on November 24, 2004. Burundi and Rwanda acceded to the Protocol in July 2007, following their accession to the East African Community Treaty. Both Burundi and Rwanda are riparians to the Kagera River, which is the largest river flowing into Lake Victoria. For more information on the Protocol, *see supra* n. 447.

[720] The Nalubaale/Kiira hydropower dam complex is about two kilometers from what used to be the Rippon Falls, which itself is about one kilometer from where the Nile exits from Lake Victoria.

Lake Victoria, and that Bujagali will simply reuse the water coming from the upstream complex.[721]

With regard to the alternative for the Bujagali Project, the Report indicated that at present, the only feasible alternative to large-scale hydropower generation in Uganda is thermal power. However, thermal power is not only more costly than hydropower, it also has negative environmental effects including air pollution, noise, potential for fuel spills, and greenhouse gas emissions.

The Report addressed the issue of the impact of climate change on Lake Victoria. It cited a number of studies analyzing climate change impact on the Lake and the Nile River hydrology which concluded that no significant reduction in hydrological flow is expected as a result of climate change during the life of the project. The Report indicated that the conclusions of the SSEA agreed with those studies, and averred that, taking into account the uncertainties associated with any prediction, climate change is likely to increase the availability of water and runoff in the Lake Victoria Basin. Climate change would therefore potentially bring upside benefits rather than downside risks to the economics of the project. The Report concluded that, given all the available evidence, there was no basis for identifying climate change as a significant risk factor for the Project.[722]

The Report also dealt with the issue of the adequacy of the analysis of cumulative impacts. It indicated that the assessment of cumulative impacts which would normally be conducted for a private sector project was expanded beyond the project area, to encompass the 320-kilometer reach of the Victoria Nile River between Lake Victoria and Lake Albert. The analysis took into account other initiatives such as the other foreseeable hydropower project, Karuma, also on the Victoria Nile, but physically separated from Bujagali by Lake Kyoga. The Report stated that the SSEA was made public in April 2007, and is referred to in the PAD, and also through a link on the Bank's Bujagali web site, to the

---

[721] The Report clarified, however, that increased demand for electricity and the drought that has engulfed the region since 1999 has affected the Lake level, and subsequently the Agreed Curve. In order to keep the basic components of the economy running, there was over-abstraction of water for power generation, and the Report attributed the decrease in the Lake level, in almost equal percentages, to the over-abstraction for power generation and the drought. Nonetheless, the Government of Uganda reduced power output from Kiira and Nalubaale dams from 265 megawatt peak capacity in 2004, to 120 megawatt capacity in 2006, in an effort to return to the Agreed Curve.

[722] More details on the project in general, and those issues in particular, are available at: http://www.worldbank.org/bujagali. It is stated there that the Bank is supporting improved environmental management of the Lake Victoria and the Nile Basin through the regional Lake Victoria Environmental Management Project (LVEMP) and the NBI. Specifically, LVEMP has assisted the riparian countries in a number of areas including better measurement of fish stocks, reduction of infestation of water hyacinth to manageable levels through biological and other controls, and study of the issue of the hydrology of the Lake and its water quality. For more information on LVEMP *see supra* n. 444.

Nile Basin web site where the SSEA can be found.[723] The Report concluded with a statement that regarding each of those areas, no action is planned beyond on-going supervision.

As discussed earlier, the Project Details attached to the notification letter usually relies on the EIA to make the determination that Bank-financed projects will not cause appreciable harm to any of the riparians.[724] The earlier discussion and the above overview of the Bujagali Project indicate clearly that the assessment of compliance by the Bank with OP 7.50, Projects on International Waterways, is partly addressed by assessment of its compliance with OP 4.01, Environmental Assessment. The process of notification itself, regarding inclusion of all the riparian states, the information furnished about the project, the time for response, as well as the responses and how the Bank will deal with them, are all undertaken under the provisions of OP 7.50. However, the content of the notification and the data and information used for making the determination that the project will not cause appreciable harm are derived largely from the EIA. This fact clearly underscores the close linkages between the two policies. Indeed, the substantive elements of projects on international waterways are basically environmental ones, even when they relate to water quantity, and accordingly are addressed through OP 4.01 on Environmental Assessment.

---

[723] The Report referred to the Bank's Bujagali web site: http://www.worldbank.org/bujagali.

[724] *See* Chapter 5, Part 5.4 of this Book.

CHAPTER **10**

Conclusion

It was in the early 1950s that the Bank was initially confronted with the complex and intricate issues concerning projects on international waters. The difficulties faced were immense—the large number of shared rivers, existing and emerging disputes over many of those rivers, and the fact that only some of the riparians were members of the Bank. That situation was further compounded by the paucity of international law principles in this field. Furthermore, few agreements on the sharing and protection of international rivers existed at that time, and almost none of them encompassed all the riparians. When the Bank started considering financing projects on international waters, the ILA had not even established its Water Resources Committee.[725] The IIL had only issued its Madrid Declaration in 1911 which was widely criticized because it required the consent of the other riparians for projects on the shared river.[726] The principle of absolute territorial integrity that the Madrid Declaration impliedly embraced has been seen as the corollary of the principle of absolute territorial sovereignty, or the Harmon Doctrine, that was widely rejected, including by its own sponsors.[727] Thus, there were no established international legal principles for the Bank to rely on or to draw from. In dealing with those issues, the Bank had to be guided by its character as an international financial cooperative institution, and by the requirement under its Articles of Agreement to act prudently in the interests of both the borrower, as well as the other members of the institution.

It took some time and intensive deliberations within the Bank, involving both the legal and operational staff, to develop some general guidelines on how to handle the international water aspects of such projects. Based on its Articles and character, the Bank determined from its early years that it would not finance projects that would cause appreciable harm to any of the other riparians. In such cases or when a dispute existed, the Bank required conclusion of an agreement between the riparians. Other projects were to be treated and dealt with on a case-by-case basis, depending on the project and the international waterway, under what was

---

[725] As may be recalled, the ILA established the Water Resources Committee in 1954, which issued the first statement in 1956. *See supra* n. 71.

[726] *See supra* n. 64.

[727] *See supra* n. 60.

then known as the "flexible approach." However, the approach did not go beyond shared rivers, nor did it address the types of projects to be financed.

The Ghab Project in Syria which was processed in the mid-1950s showed clearly that the challenges of projects on international waters were far more intricate and complex than was initially thought. The Bank was faced with the first objection from one of the riparians to one of the projects proposed for Bank financing. The objection brought to a head the competing interests and demands of the riparians of the river where the project would be carried out. However, concomitant to those challenges, the objection provided the Bank in 1956 with the opportunity to issue its first directive, OM 8, for Projects on Inland International Waterways, less than 10 years after it started its operational work. The importance and significance that the Bank attached to the matter was underscored by the fact that the Memorandum was number 8, preceded only by few others, and was the first to address a substantive operational matter. It is worth noting that it took close to 30 more years before the Bank would issue its first policy on the environment.

OM 8 established an early warning system by requiring the Bank staff to bring to the attention of senior management any project on an international waterway, together with a recommendation of procedures for dealing with the international aspects of the project. The OM thus codified the flexible approach that the Bank had been following, and went further, defining the international waters and projects covered therein. However, the scope of such international waterways and projects kept evolving and expanding through the years, based largely on the Bank implementation experience.

Indeed, the policy is characterized by a paced evolution in a number of aspects. It started with a short OM in 1956, to which a number of new provisions were added in 1965, with the current parameters of the policy arrived at, and approved by the Executive Directors, and issued in 1985 as OMS 2.32. A large number of gaps have been filled, and ambiguities clarified, through the implementation experience of the policy during the last quarter of a century.

The evolution covered a number of aspects. First, the 1985 OMS went beyond the notification requirement to codify the facilitative role of the Bank and to confirm that cooperation and goodwill of riparians is essential for the efficient utilization and protection of the waterway. The policy states that the Bank stands ready to assist the riparians in establishing collaborative arrangements for the entire waterway or any part thereof. This proactive role has been pursued through a number of initiatives, projects and programs in many basins, and has had positive results in a number of them.

Second, inland waters under that OM covered not just international rivers, but included lakes and canals. The expansion of the waters covered under the policy has been both vertical as well as horizontal. The policy gradually extended horizontally from rivers, lakes and canals, to include any tributaries to or components of those waterways. This was further expanded to include semi-enclosed coastal waters as well as closed seas. Moreover, application of the policy was also extended to national rivers flowing into such closed seas and semi-enclosed coastal waters. This international dimension of national rivers finds support, as discussed earlier, in the 1982 Convention on the Law of the Sea.[728] The addition of semi-enclosed waters to the policy in 1965 was necessitated by the absence of any Bank directive on the environment at that time. Those types of waters continue to be covered by the policy because the environment directives issued since that time do not include a notification requirement. Thus, the policy continues with this unusual feature of mixing fresh waters and sea waters.

Moreover, since 1990 the Bank, through practice and precedents, has included transboundary groundwater in its definition of international waterways, thus expanding the definition vertically. This inclusion was prompted by the Sahara Regional Development Project in Algeria which shares the aquifer where the project would be carried out with Tunisia.[729] As may be recalled, the issuance of OM 8 in 1956 was prompted by the objection of Turkey to the Ghab Project in Syria, and the issuance of OMS 2.32 in 1985 was the result of the objection of Iran to the Igdir-Aksu Project in Turkey. The inclusion of transboundary groundwater as part of the policy was reconfirmed less than 10 years after the Algeria Project in the Disi Project in Jordan. This practice is now firmly established notwithstanding the absence of an explicit reference to transboundary groundwater in the policy. The Bank even deliberated whether a local aquifer connected to an international river should be considered as a transboundary aquifer, and covered by the precedents and practice based on the policy. Clearly, the Middle East Region has played an important role in the evolution of the Bank policy for projects on international waterways.[730]

One of the aspects of the policy worth noting relates to the riparian states whose interests and concerns need to be taken into account. The Articles of Agreement of the Bank have directed the Bank to act prudently in the interests of

---

[728] *See* Redgwell, *supra* n. 288.

[729] As explained earlier, *supra* n. 637, it has been established since the mid 1990s that Libya also shares this aquifer, now known as the North-Western Sahara Aquifer System, with Algeria and Tunisia.

[730] For a detailed analysis of the challenges in connection with the management and sharing of the major watercourses in the Middle East, *see* Kolars, *supra* n. 505. *See also* Arnon Soffer, *Rivers of Fire: The Conflict of Water in the Middle East* (Rowman & Littlefield Publishers 1999); Daniel Hillel, *Rivers of Eden—The Struggle for Water and the Quest for Peace in the Middle East* (Oxford University Press 1994); and Shapland, *supra* n. 47. *See also supra* n. 153. The term "Middle East region" as used in this Book, and the above quoted books, includes the North African countries.

both the particular member in whose territories the project is located, as well as the members as a whole. However, from the very beginning the Bank took the position that, consistent with the principles of international law, affected riparian states need not be members of the Bank to be covered by its policy. Accordingly, OM 8 instructed the staff to ensure that the Bank's concerns should go beyond members, to include non-members as well. The emphasis of non-members' interests and rights over the shared waterway reflects basic principles of international law which the Bank could not possibly overlook. Dispute avoidance, fairness and equity towards, and equality of all the riparians under international law, prompted the Bank to look beyond its member countries. This emphasis of the rights of riparian non-members of the Bank continues until today,[731] despite the fact that only a few countries are still not members of the Bank.[732]

However, the due diligence that the Bank exercises with regard to the interests of non-members is limited to ensuring that the proposed project will not cause appreciable harm to such non-member riparians. The Bank also ensures that any agreements in place are observed, and that the proposed project would not contravene the provisions of such agreements. The due diligence stops there and has not included notification of non-member states for projects on international waterways they share with other members.

Along the same lines, the notification process under the policy includes both developing member countries as well as developed member countries. Projects in Poland on the Odra (Oder) River which Poland shares with Germany and the Czech Republic require notification of both countries.[733] Similarly, Greece was notified of a project in Bulgaria on one of the tributaries of the Maritsa River which flows into Greece.[734] Notification for Bank-financed projects to the Danube Commission under the Danube Convention, the Black Sea Commission under the Bucharest Convention, or UNEP under the Mediterranean Convention[735] would involve some of the developing member countries of the Bank in Europe by the respective Commission, or by UNEP. Similarly, projects on the

---

[731] Reference to non-members is included in paragraphs 1(a) and 3 of OP 7.50.

[732] As of January 2009, the member states of the United Nations were 192, while IBRD had 185 member countries. For the list of the UN members *see*: http://un.org/members/list.shtml. For the membership of the World Bank *see* the World Bank web site, *supra* n. 1. *See* also *The World Bank Annual Report* 2008, *supra* n. 1. For the gradual growth in IBRD's membership *see supra* n. 211. The UN members which are not IBRD members are Andorra, Cuba, Liechtenstein, Monaco, Nauru, Democratic People's Republic of Korea (North Korea), and Tuvalu.

[733] *See* Poland, Odra River Basin Flood Protection Project, *supra* n. 419.

[734] *See* Bulgaria, Municipal Infrastructure Development Project, *supra* n. 415.

[735] *See supra* n. 417.

Colorado River or the Rio Grande in Mexico would require notification of the United States of America, since the two countries share those rivers, if the projects do not fall under one of the exceptions to the notification requirement of the policy. Furthermore, the sharing of those rivers is regulated by a treaty which the Bank will need to take into account when financing projects on those rivers.[736] The involvement of both developing and developed member countries of the Bank is another unique feature related to the Bank policy for projects on international waterways.

Another important, and indeed significant, feature of the policy is the requirement that all riparians, both downstream as well as upstream, be notified of the project and given the opportunity to respond to the notification with any comments they may have. The requirement that notification includes both downstream as well as upstream riparians stems from the fact that a project in any riparian state can affect the others, regardless of their location on the shared waterway. As indicated earlier, the downstream riparians can be affected by the physical impacts of water quantity and quality changes caused by water use by the upstream riparians. On the other hand, the upstream riparians can be affected by the potential foreclosure of their future use of water caused by the prior use of water, and the claiming of rights to such waters by the downstream riparians. The notification letters and Project Details prepared by the Bank, or by the borrowers with the Bank assistance, have helped considerably in explaining how projects can affect both groups of riparians, and in clarifying the concept of potential foreclosure of future uses.

By requiring notification to both downstream as well as upstream riparians, the Bank policy contributes substantially to correcting the misconception that only upstream riparians can harm downstream riparians. Thus, the policy has clarified through practical application that projects in one riparian state can affect the other riparian states, regardless of the location of those states. In so doing, the policy also establishes the linkages between the obligation not to cause harm and the principle of equitable and reasonable utilization. When the Project Details attached to the notification letter specify the amount of water to be abstracted from the shared river, lake or aquifer for the project, and makes the determination that the project will not cause appreciable harm to other riparians, the notification is implicitly stating that this amount of water is within the reasonable and equitable share of that riparian. Similarly, when a project is being processed in a downstream country, it could be objected to by other upstream riparians as giving more water rights to that riparian than what is equitable and reasonable. Those

---

[736] *See supra* n. 60 for the Treaty on the Colorado River and Rio Grande between the United States of America and Mexico.

water rights, it could be alleged, could be claimed as established rights, or *fait accompli*, thus foreclosing their future uses, and precluding other riparians from using that amount of water. As discussed earlier, foreclosure of future uses was one of the reasons for the objection of Ethiopia to the Baardhere Project in Somalia. It should also be recalled that Ethiopia is the upstream riparian for the Juba River where the Baardhere Project was supposed to be constructed.[737] As noted before, there has been a gradual comprehension of the concept of the foreclosure of future uses by some states, as reflected in the Senegal Water Charter.[738]

Thus, this practical application of the notification process of the policy shows the clear linkages between the principle of equitable and reasonable utilization, and the obligation not to cause harm, and indeed the confluence of the two principles. It is true that the ILA Helsinki Rules and the UN Watercourses Convention both require notification of other watercourse states, without limiting the requirement to downstream riparians. However, the Bank policy has clarified this matter through the practical application of the requirement of notification of all the riparians, and through providing such riparians with the opportunity to determine the effects of the project on their interests on the shared waterway. This is, no doubt, a significant and major contribution of the policy to the evolution and progressive development of international water law.

The policy has also operationalized the process of notification through the Bank or another third party, on behalf of the borrower, if the borrower does not wish, for one or another reason, to undertake the notification itself. By undertaking the notification on behalf of the borrower, the Bank becomes the channel of communication between the riparians, and would thus helps fill a diplomatic vacuum. However, if the borrower does not wish to undertake the notification itself, and will not allow the Bank to do so, then the Bank discontinues processing of the project. This is a fair and logical position given the character of the Bank as an international financial cooperative institution. Indeed, all riparians have been or will be, at one point in time, beneficiaries of the policy when notified of projects on waterways they share with the other riparians, provided with the Project Details thereon, and given an opportunity to comment. Good neighborliness and reciprocity should be adequate reasons for persuading reluctant riparians of the reasonableness of the requirement of notification, if not by them, then by the Bank on their behalf.

The Bank carried the procedure of notification through a third party further and allowed the NBI Secretariat to undertake the notification on behalf of, and at the request of any of the 10 Nile riparian countries. This process has now evolved

---

[737] *See* Somalia, Baardhere Dam Project, *supra* n. 507.

[738] *See supra* n. 392.

to become a regular method, and is assisting in strengthening the NBI Secretariat itself.[739] The UN Watercourses Convention adopted a similar approach, requiring the watercourse states to undertake notification and exchange of data and information through any indirect procedures accepted by them when there are serious obstacles to direct contacts between them.[740] The NBI process provides a good precedent for such indirect procedures.

Furthermore, the Bank recognizes and follows existing arrangements between the riparian states, including those dealing with notification. River basin organizations and commissions that are duly established and given authority to receive such notification under their respective legal instrument would be notified of Bank-financed projects on the shared waterway. Such organizations include the LCBC, the NBA, the ICPDR, the ICPOAP, the Administrative Commission, the Consultation Mechanism under the OSS, as well as UNEP, which is not even a river basin organization, but is designated by the Mediterranean Convention to receive such notifications.

The Bank also adheres to existing agreements when they lay down strict rules for notification. If the Bank determines that the project falls under one of the exceptions to the notification requirements of the Bank policy, but an existing agreement requires notification or exchange of data and information for such types of projects with the other riparians, then the Bank would seek to secure compliance with this agreement. Conversely, if the agreement requires notification only for projects that may cause significant adverse effects, and the project is only expected to have minimal effects, the Bank, nevertheless, would require notification under its policy. This approach underscores the caution that the Bank exercises with regard to projects on international waterways, and its determination to clearly convey to the riparians that it will not finance a project that may cause significant adverse effects or appreciable harm to any riparian.

Despite the repeated references to the terms "appreciable harm" and "adverse effects" none of the Bank directives issued since 1956 defined these terms. The ILC defined the term "appreciable harm" in a general way,[741] and the Sixth Committee did the same with the term "significant adverse effects."[742] The Bank uses its Environment Assessment policy to fill this gap, and the environment impact assessment provides the basis for making the determination with regard to the degree of harm or adverse effects that the project may cause. Reliance on OP 4.01

---

[739] It may be recalled that in a project in Kosovo, notification was undertaken by UNMIK, *see supra* n. 385.

[740] *See* Article 30 of the UN Watercourses Convention, *supra* n. 180.

[741] *See supra* n. 345.

[742] *See supra* n. 371.

would take place even in situations where the issues relate purely to the quantity of water, abstracted for the project, and not directly to quality of such waters. The Project Details need to rely on the EIA to show that the project will not cause appreciable harm. Along these lines, the request for additional information from one or more of the notified riparians usually centers around additional details on the environmental impacts of the project. This interface of the two policies was clearly manifested in the Inspection Panel Investigation Report on the Bujagali Project, as well as the Bank Report and Recommendation thereon.[743]

Another interesting interface of the Bank policy for projects on international waterways is with the policy for projects in disputed areas. These areas can be land, or water, as reflected in the case of the Caspian Sea. They can also relate to rivers or lakes as the case with the Orange River or Lake Malawi/Nyasa. In these cases, the issues are more complex because they relate to sovereignty and ownership, whereas in the case of international waterways, the matter usually relates to the share of each riparian of the waters of the waterway in question. This clearly makes the interface between the policy for projects on international waterways and that for projects in disputed areas a complex and intricate one. Yet, the synergies between the two policies are manifest, particularly as one notification letter can be used to address both matters.

It is worth noting that the implementation experience of the policy for projects on international waterways, and the varying types of responses of the notified riparians, has generated a vast and interesting literature, and highlighted a number of issues that were not conceived by the original drafters of the policy. The distinction between a response that could be characterized as a "qualified no-objection" and the one that is considered a "conditional no-objection" may not be easy to draw, and would require extra caution on the part of the Bank. Interestingly, neither type of response has been specified in the policy, and in both cases the process of identifying and dealing with them has been *ad hoc*, and evolutionary.

All the other instruments discussed earlier establish a threshold for notification, such as "significant adverse effects." Projects falling below the threshold under those instruments would not require notification. However, the Bank did not establish such a threshold because it does not finance projects that may cause appreciable harm to other riparians. If the Bank were to follow this approach and establish significant adverse effects as the threshold for notification, then the notified riparians would simply rely on this determination and object to the project. Moreover, adoption of such a threshold would arguably be inconsistent with the Bank's Articles of Agreement, specifically the requirement to act prudently in

---

[743] *See supra* n. 703.

the interests of all the riparians. In this connection the Bank seems to follow closely the *ratio decidendi* of the *Lake Lanoux* arbitral award, which stated that it is the other riparian, and not the state planning an activity, that should decide whether its interests are affected, and should thus be provided with information on the proposed project to enable it to make that decision. Yet, the views of the notified state are not conclusive, and the Tribunal itself indicated that such state does not have veto power over the proposed project.[744]

Consequently, the Bank policy provides for three exceptions to the notification requirement. This is another unusual feature of the Bank policy, and stems from the absence of a threshold for notification. The first exception, dealing with the rehabilitation of existing schemes, has been widely used, sometimes because the project actually consists of such components, but in other times simply to avoid notification, either by the borrower, or by the Bank. This latter approach is unfortunate given that none of the riparians has veto power over the project. Limiting the project to rehabilitation activities when other new activities could have been financed forgoes, unnecessarily and unfortunately, development opportunities that may not recur.

With regard to the second exception which deals with water resources surveys and feasibility studies, it is worth noting that other riparians have been consulted on some of the studies that are of significant magnitude. Furthermore, information on some of those studies has been widely disseminated. The Bank reached the conclusion regarding the need for such consultations or disclosure of information because of the magnitude and visibility of these studies, as well as the wider objective of dispute avoidance.

The third exception, dealing with projects in a tributary originating in the lowest downstream riparian, is unfortunate in a number of ways, and has proven unnecessary. It is unfortunate because it fails to view the river and its tributaries as a single hydrologic unit where any consumptive use would affect overall water availability in the basin and foreclose the future uses of such amounts of water. It may be recalled that the term "downstream riparians" was dropped from the 1989 OD, and replaced by "other riparians." Yet, five years later, the term was used in the 1994 policy when this exception was codified. In addition to being unfortunate, the exception has proven unnecessary because it has hardly been used since it was codified in 1994.[745]

---

[744] *See supra* n. 356.

[745] As discussed earlier, the processing of the project that prompted this exception (Burma, Second Tank Irrigation, 1988) was discontinued, *see supra* n. 585. The other project, the Nigeria, Local Empowerment and Environmental Management Project was processed under the Niger Basin Convention, *see supra* n. 586.

A number of criticisms have been directed towards the policy. One such criticism is that the policy is not concerned with equitable and reasonable utilization which is the guiding principle of international water law. As indicated earlier, the policy could have been crafted around this principle. However, that would have required an adjudicative role for the Bank each time it finances a project on an international waterway. Such a role would require the Bank to decide whether the particular use under the project falls within the equitable and reasonable share of that state, which in turn would require making a determination on the allocation of the waters of the entire river. This decision would obviously have to be made by the riparians themselves, and not by the Bank. Moreover, even if the Bank were to make that determination, it would most likely be challenged by one or more of the riparians, derailing the whole process of financing a specific project into a larger and more complex issue of the allocation of waters; a role that the Bank cannot, and indeed should not, exercise. Furthermore, the notification requirement of all the other international instruments in this field—the ILA and the IIL Rules, the UN Watercourses Convention as well as other regional and bilateral instruments on shared watercourses—is based on the threshold of significant adverse or material effects. None of those instruments has established equitable and reasonable utilization as the basis for notification, and it is not clear why the Bank should do that. In fact, as discussed and explained throughout this Book, the practical application of the requirement that the Bank would not finance a project that causes appreciable harm to other riparians has meant, for all practical purposes, taking into account the principle of equitable and reasonable utilization.

Another criticism is that the policy favors the downstream riparians because they are the ones who would be harmed by activities of the upstream riparians and not *vice versa*. The discussion throughout this Book has explained that the requirement of notification goes both ways, and the reasons for that have been amply discussed and explained.

It is also contended that the policy is pro-rich developing countries, or the middle-income countries. This is because those countries can carry out projects on international waterways which may even cause appreciable harm to other riparians without notifying them, because they can fund those projects from their own resources. In this way they do not need to comply with the notification requirement. On the other hand, the poor countries that do not have the resources to develop their shared waterways need the Bank to assist them in financing such projects, and as such are subject to the notification requirement under the Bank policy. While the facts of this argument are true, it has been emphasized throughout this Book that the notified states have no veto power over the project. Even when they object, the Bank can proceed with financing the project if it can satisfy itself and its Executive Directors that the project will not cause appreciable harm to any of the other riparians. In the only case when the Bank has appointed independent experts, those experts confirmed the Bank's determination that the project would not cause appreciable harm to the other riparians. In fact the Bank's approach to objections by other riparians has influenced the thinking in this area, as evidenced by the UN Watercourses Convention and other instruments dealing with international watercourses. The Convention calls for the appointment of a fact finding commission in case of a dispute, and requires the parties to consider the findings of the commission in good faith.[746]

The fact that the Bank has never sustained any objection by one of the riparians to any of its financed projects has raised questions as to whether the Bank policy is about notification, or is simply about exchange of data and information on the proposed project. It is true that no objection by a notified riparian to a Bank-financed project on an international waterway has been sustained. However, this conclusion is usually reached after a thorough analysis of the reasons for such objections by a number of units within the Bank. As stated repeatedly, one main philosophical foundation of the policy is that the Bank, *ab initio*, will not finance a project that is expected to cause appreciable harm to any of the riparians. Hence, notification is not undertaken because the project will cause appreciable harm to other riparians. Rather, it is undertaken to confirm to the other riparians that the Bank-financed projects will not cause appreciable harm to them. In this way it can be argued that the notified riparians have no veto power over the project as long as the Bank adheres to its guiding principle of not financing a project that will cause appreciable harm to other riparians.

It has also been alleged that the policy has not been applied consistently and systematically, and that some riparians have been treated more favorably than other riparians. No doubt there have been oversights, and even cases of omission, with regard to the application of the policy and the notification requirement. Such oversights and omissions have occurred in the application of most Bank policies in the past. However, since 1997 when the safeguard policies were carved out and highlighted, and the need for full compliance emphasized, there has been, by and large, consistent and systematic application of the policy across all projects and regions, and for all riparians. When some of the borrowers have objected to notification, the processing of those projects has either been discontinued, or the projects have been redesigned to remove the components on the shared waterway, provided that any such a project remains viable and coherent.

Another criticism voiced against the policy is its strict requirement of notification even when the project would have minimal or no effects on other riparians. Such projects typically include small-size waste-water treatment projects or

---

[746] *See* Article 33 of the UN Watercourses Convention, *supra* n. 180.

components, community-driven development and social funds projects with small water supply or irrigation components, and small-size run-of-river hydropower projects with limited or no storage. They also include minor flood protection works, and a wide range of sector projects with small water components that have little or no effects on the quantity or quality of water flows to other riparians. Concerns have been voiced by both borrowers and Bank staff that when effects are minimal, or when there are no effects, notification poses an unnecessary burden on the borrowers and the Bank, and can result in delays in project processing. In addition, any objection from one or more of the riparians to any of these types of projects will most likely not be sustained by the Bank. This, no doubt, is a valid criticism, and needs to be addressed, without compromising the basic philosophical and jurisprudential foundation of the policy emanating from the Bank's Articles of Agreement and its character as an international financial cooperative institution.[747]

The fact that the policy does not include an explicit reference to transboundary groundwater is another criticism of the policy, especially by those who are not aware of the application of the policy to transboundary groundwater through precedents and practice since 1990. This is true among a number of scholars, borrowers, and even Bank staff. An explicit inclusion of transboundary groundwater in the Bank policy is certainly needed to codify the Bank practice since 1990, and to align the policy with other international legal instruments in this field.

Thus, none of those criticisms manifests a major flaw in the policy, or its evolution, context, or application. In fact, despite the complexity of the issues, and the paucity of the principles of international law in the field of shared watercourses in the middle of the last century, the Bank was able to adopt its first policy for projects on international waterways in 1956. The policy has evolved over the years, and the vast implementation experience has provided sound and important basis for its updating and refinement, as well as for filling its gaps. It now includes a number of features, discussed throughout this Book and overviewed in this Chapter, which make it a unique instrument. Indeed, it stands out as the only international instrument dealing with shared waterways that is a fairly comprehensive one, and with a global reach and practical application. Although it is based on the existing principles of customary international water law, the policy has made major contributions to the evolution and progressive development of such principles, and will continue to do so.

---

[747] One possible approach for addressing this matter is to have one exception to the notification requirement under the policy that covers projects with no effects or minimal effects on other riparians. This approach would require a clear definition of the term "minimal" and transparent and credible procedures for making such a determination.

APPENDIX **1**

OM No. 8
March 6, 1956

OPERATIONAL MEMORANDUM

PROJECTS ON INTERNATIONAL
INLAND WATERWAYS

1. Loan projects on international inland waterways are likely to give rise to problems which go far beyond the usual problems of project analysis. They may affect relations not only between the Bank and its borrower but also among governments. These projects therefore need special handling and it is vital that the Bank decide how each case will be handled before discussions have reached a stage which it would be embarrassing to introduce new questions.

2. The following procedure should therefore be followed:

(a) Whenever a hydroelectric, irrigation, flood control, navigation, drainage or similar project involves use of a river, canal, lake or other inland waterway, the first inquiry should be whether the waterway forms a boundary between, or flows through, two or more countries, whether members or non-members of the Bank.

(b) If it does, the management should be informed promptly and the Working Party for the project and the interested department should make it their first order of business to propose and obtain management approval of a procedure for dealing with the international aspects of the project.

(c) No steps should be taken to investigate the merits of the project or to process the project without prior approval from the management. This requirement is to be observed even if the international aspects appear to be covered by international agreement or if there appears to be *prima facie* evidence that the project cannot be adversely affected by, or will not adversely affect, up-stream or down-stream riparian states.

OM No. 5.05
January 1, 1965

OPERATIONAL MEMORANDUM

PROJECTS ON INTERNATIONAL WATERS

1. Projects on International Inland Waterways

(a) Projects on international inland waterways are likely to give rise to problems which go beyond the usual problems of project analysis. They may affect relations not only between the Bank/IDA and the borrower but also between governments. These projects therefore need special handling and it is vital that the Bank/IDA decide how each case will be handled before discussions have reached a stage at which it would be embarrassing to introduce new questions.

(b) The following procedure should therefore be followed:

(i) Whenever a hydroelectric, irrigation, flood control, navigation, drainage, sewage, industrial or similar project involves use or pollution of a river, canal, lake or other inland waterway, the first inquiry should be whether the waterway forms a boundary between, or flows through, two or more countries, whether members of the Bank/IDA or not.

(ii) If it does, the matter should be brought promptly to the attention of the Chairman of the Loan Committee. Before any steps are taken to investigate the merits of the project, the Area Department concerned should propose to the Chairman of the Loan Committee and obtain his approval to a procedure for dealing with the international aspects of the project. This requirement is to be observed even if the international aspects appear to be covered by international agreement or if there appears to be *prima facie* evidence that the project cannot be adversely affect by, or will not adversely affect, up-stream or down-stream riparian states.

2. Projects Involving other International Waters

The procedures set forth above should be followed also in the case of any project involving such international waters as bays, gulfs, straits, or channels bound by several states or, if within one state, recognized as necessary channels of communication between the open sea and other states.

3.   Presentation of Loans/Credits to the Executive Directors

When presenting loan or credit projects on international waters to the Executive Directors both the Appraisal Report and the Report and Recommendation of the President should state that the Bank/IDA has considered the international aspects of the project and is satisfied that: (i) the issues involved are covered by appropriate arrangements between the borrower and other riparians; or (ii) the other riparians have stated (to the borrower or to the Bank/IDA) that they have no objection to the project; or (iii) the project is not harmful to the interests of other riparians and their absence of express consent is immaterial or their objections are not justified.

APPENDIX 3

OMS No. 2.32
April 1985

OPERATIONAL MANUAL STATEMENT

PROJECTS ON INTERNATIONAL WATERWAYS

Basic Policy Approach

1. Projects on international waterways require special handling as they may affect relations not only between the Bank[1] and its borrowers but also between States, whether members of the Bank or not. In this connection, the Bank recognizes that the cooperation and goodwill of riparians is essential to the most efficient utilization and exploitation of international waterways for developmental purposes. The Bank, therefore, attaches the utmost importance to riparians entering into appropriate agreements or arrangements for such utilization for the entire waterway system or any part thereof, and the Bank stands ready to assist riparians to this end. In cases where differences remain unresolved, the Bank, prior to financing a project will normally urge the State proposing a project to offer negotiations in good faith with other riparian(s) with a view to reaching appropriate agreements or arrangements.

Applicability

2. This Operational Manual Statement (OMS) covers the following:

(a) Types of international waterways:

(i) river, canal, lake or any similar body of water which forms a boundary between, or any river or body of surface water which flows through, two or more countries whether members of the Bank or not;

(ii) any tributary or any other body of surface water which is a part or a component of any waterway described in (i) above; and

(iii) bays, gulfs, straits or channels, bounded by two or more States whether members of the Bank or not, or, if within one State recognized

---

[1] All references in this Statement to the Bank and to Bank Loans include IDA and IDA credits.

241

as necessary channels of communication between the open sea and other States, and any river flowing into such waters.

(b) Types of projects:

(i) hydroelectric, irrigation, flood control, navigation, drainage, water and sewerage, industrial or similar projects which involve the use or pollution of international waterways as described above; and

(ii) detailed design and engineering studies of projects under (i) above, including those to be carried out by the Bank as executing agency.

General Guidelines

3. The procedures contained in the following paragraphs are intended to ensure that the international aspects of a project on an international waterway are brought to the fore and dealt with at the earliest possible opportunity.

4. The presence of any potential international water rights issue should be ascertained as early as possible during project preparation and be described in the Project Brief. Before an appraisal is undertaken for a proposed project on an international waterway, a memorandum should be addressed to the Chairman of the Loan Committee giving all relevant information on the international aspects of the project, and the Chairman of the Loan Committee should be informed promptly during the project cycle of any significant event which may have taken place in connection therewith.

Notification

5. As early as possible during the identification stage of the project cycle (see OMS 2.00, paras. 7–14) the Bank should advise the State proposing the project on an international waterway (hereinafter called "beneficiary State") that, if it has not already done so, it should formally notify the other riparian(s) of the proposed project. However, if the beneficiary State indicates to the Bank that it does not wish to give notification, the Bank will normally give such notification to the other riparian(s). If the beneficiary State does not wish to give notification and objects to notification proposed by the Bank, the Bank will discontinue further processing of the project. The Executive Directors concerned should be informed of these developments and of further steps in connection therewith.

6. The notification should contain to the extent available sufficient technical and other necessary specifications, information and data (hereinafter called "Project Details") to enable the other riparian(s) to determine as accurately as possible the potential for appreciable harm by the proposed project by way of deprivation of water rights, pollution or otherwise. If the Project Details are not

available at the time of notification, the same should be made available to the riparian(s) as soon as possible after the notification. The Bank staff should be satisfied that the Project Details are adequate for the purposes of making such determination. If it is considered that the Project Details will not be available prior to appraisal, the proposed project should be brought to the attention of the Chairman of the Loan Committee giving all relevant facts related to the international aspects of the project, and seeking approval to proceed with appraisal.

7. The other riparian(s) should be allowed a reasonable period of time, which should normally not exceed six months from the dispatch of the Project Details, to communicate its response to the beneficiary State or the Bank, as applicable.

8. The Bank staff should also make efforts to ascertain whether the riparians of the international waterways concerned have established any institutional framework with respect to such waterways and if so, the scope of activities and functions of such institution as well as the status of its involvement in the proposed project, with particular regard to any notification which may be required to be given to it.

9. It should be noted, however, that notification to riparians will not be required:

(a) Where in case of projects which involve additions or alterations by way of rehabilitation, construction or otherwise to any existing works or ongoing schemes, it is the judgment of the Bank that such projects will not adversely change the quality or quantity of water flows or that such projects will not be adversely affected by the uses of water that the other riparian(s) might make. However, if there is any agreement or arrangement between the riparians, the Bank staff should make efforts to secure compliance with the requirements of such agreement or arrangement.

(b) In case of water resource surveys and feasibility studies on, or involving, international waterways. Beneficiary States for such projects should, however, be required to include in the terms of reference for such surveys and studies an examination of any potential riparian issues.

Response/Objections

10. If the beneficiary State or the Bank, after giving the notice, receives from the other riparian(s) a positive response which may be in the form of, for example, consent, no objection, support to the project or a confirmation that the project will not be harmful to its interests, or the other riparian(s) has (have) not responded at all within the stipulated time-limit, the Regional Office should, in consultation with the Legal Department and the other departments concerned, address a memorandum to the Chairman of the Loan Committee giving all relevant facts including Bank staff assessment of whether the project will cause

appreciable harm to the interests of a riparian(s) or be so harmed by the uses of water that the other riparian(s) might make, and seeking approval for the further action to be taken.

11. If the other riparian(s) has (have) raised objections to the proposed project, the Regional Office, in consultation with the Legal Department and other departments concerned, should address a memorandum to the Loan Committee detailing, *inter alia*, as applicable:

  (a) The nature of the riparian issue or issues;

  (b) The Bank staff assessment of the objections, including its reasons and supporting data, as appropriate;

  (c) The Bank staff assessment of whether the proposed project will cause appreciable harm to the interests of a riparian(s) or be so harmed by the uses of water that the other riparian(s) might make;

  (d) Whether the circumstances of the case require that the Bank should, before taking any further action, urge the parties to resolve the issues through such amicable means as consultations, negotiations, good offices, etc., which step will be normally followed in the cases where objection by the other riparian(s) is substantiated;

  (e) Whether the objections are of such a nature that it would be advisable to obtain an additional opinion from independent experts in accordance with the provisions of Annex A hereto.

12. The outcome of actions taken as a result of the Loan Committee instructions will be reported to the Loan Committee by the Regional Office, in consultation with the Legal Department and other departments concerned, with appropriate recommendations for further processing of the project.

13. In case independent experts are asked to give their opinion, their conclusions will be reviewed by the Loan Committee before making a decision on whether or not to proceed with further processing of the project.

Presentation of Loans to the Executive Directors

14. The Staff Appraisal Report and the Report and Recommendation of the President for every project on international waterways should deal with the international aspects of the project and should state that the Bank staff have considered such aspects of the project and are satisfied that:

  (a) The issues involved are covered by appropriate agreement or arrangements between the beneficiary State and other riparian(s); or

  (b) The other riparian(s) has (have) given a positive response to the beneficiary State or to the Bank in the form of, as applicable, consent, no objection, support to the project or confirmation that the project will not be harmful to its interest; or

(c) In all other cases, in the assessment of the Bank staff the project, as proposed, would not cause appreciable harm to the other riparian(s), or would not be so harmed by the use of waters by other riparian(s). The Report and Recommendation of the President should also contain the salient features of objections, and, where applicable, the report and conclusions of the independent experts.

Annex A

TECHNICAL ADVICE OF INDEPENDENT EXPERTS

1. Whenever it is decided by the Loan Committee to seek an additional opinion in accordance with the provisions of this Statement, the Chairman of the Loan Committee will request the Vice President, Operations Policy Staff (VPOPS), to seek the advice of one or more independent experts for this purpose.

2. The VPOPS will maintain, in consultation with the Regional Offices and the Legal Department, a roster of ten independent and highly qualified experts. The roster will be brought up-to-date annually and circulated to the Loan Committee.

3. The VPOPS will, when so requested, select the experts from the roster in consultation with the Regional Office concerned and the Legal Department to ensure the appropriate expertise and terms of reference. The experts should not be nationals of any of the riparians of the river or the rivers or of other waterways involved and should have no conflict of interest in the matter.

4. The staff will provide the experts with all necessary background information and any assistance required for the efficient completion of their work.

5. The experts shall examine the Project Details. If they deem it necessary to verify the Project Details and to take any action incidental thereto, the Bank will use its best efforts to facilitate the same. Upon completion of their work, they will furnish their report and conclusions to the VPOPS for submission to the Chairman of the Loan Committee.

6. The experts will meet on an *ad hoc* basis until they submit their report. The VPOPS may ask the experts to provide an explanation of clarification of any point or aspect of their report.

7. The experts shall not have any decision-making role with respect to the processing of a project. Their technical opinion will be for purposes of the Bank only and shall not in any way determine the rights and obligations of riparians.

APPENDIX **4**

OD 7.50
April 1990

OPERATIONAL DIRECTIVE

PROJECTS ON INTERNATIONAL WATERWAYS

Basic Policy

1. Projects on international waterways require special handling as they may affect relations not only between the Bank[1] and its borrowers[2] but also between states, whether members of the Bank or not. The Bank recognizes that the cooperation and goodwill of riparians is essential to the most efficient utilization and exploitation of international waterways for development purposes. The Bank, therefore, attaches the utmost importance to riparians entering into appropriate agreements or arrangements for the efficient utilization of the entire waterway system or any part of it, and stands ready to assist in achieving this end. In cases where differences remain unresolved, the Bank, prior to financing the project, will normally urge the state proposing the project to offer to negotiate in good faith with other riparians to reach appropriate agreements or arrangements.

Applicability

2. This directive covers the following:

   (a) Types of international waterways:

   (i) river, canal, lake, or any similar body of water which forms a boundary between, or any river or body of surface water which flows through two or more states, whether members of the Bank or not.

   (ii) any tributary or any other body of surface water which is a part or a component of any waterway described in (i) above; and

   (iii) bays, gulfs, straits, or channels—bounded by two or more states, or if within one state, recognized as necessary channels of communication

---

[1] "Bank" includes IDA, and "loans" includes credits.

[2] The term "borrowers" refers to the member country in whose territories the project is carried out, whether this be the borrower or the guarantor.

between the open sea and other states—and any river flowing into such waters.

(b) Types of projects:

(i) hydroelectric, irrigation, flood control, navigation, drainage, water and sewerage, industrial, or similar projects which involve the use or pollution of international waterways as described above; and

(ii) detailed design and engineering studies of projects under (b)(i) above, including those to be carried out by the Bank as executing agency.

General Guidelines

3. The procedures described in the following paragraphs are intended to ensure that the international aspects of a project on an international waterway are brought to the fore and dealt with at the earliest possible opportunity.

4. The presence of any potential international water rights issue should be ascertained as early as possible during project preparation, and described in all project documents starting with the Initial Executive Project Summary. The Senior Vice President, Operations (OPNSV), should be kept informed throughout the project cycle by the Country Department (CD) director, through the Regional Vice President (RVP) and in consultation with the Legal Department, of such water rights issues and significant related events. Before an appraisal is undertaken for a project on an international waterway, the transmittal memorandum for the Final Executive Project Summary, addressed to the RVP and copied to the OPNSV and the Vice President and General Counsel, should be prepared in close collaboration with the Legal Department, and convey all relevant information on the international aspects of the project.

Notification

5. As early as possible during the identification stage of the project cycle (see Circular Op 87/03, filed as OMS. 2.00, *Procedures for Processing Investment Loans and Credits*, to be reissued as OD 9.00, *Processing and Documentation for Investment Lending),* the Bank should advise the state proposing the project on an international waterway (the beneficiary state) that, if it has not already done so, it should formally notify the other riparians of the proposed project. However, if the beneficiary state indicates to the Bank that it does not wish to give notification, the Bank will normally give such notification to the other riparians. If the beneficiary state does not wish to give notification and objects to notification by the Bank, the Bank will discontinue further processing of the project. The executive directors concerned should be informed of these developments and of any further steps taken.

6. The notification should contain, to the extent available, sufficient technical and other necessary specifications, information, and data (Project Details) to enable the other riparians to determine as accurately as possible the potential for appreciable harm by the proposed project through the deprivation of water, pollution, or otherwise. (If the Project Details are not available at the time of notification, they should be made available to the riparians as soon as possible after the notification.) Bank staff should be satisfied that the Project Details are adequate for the purpose of making such a determination. If, in exceptional circumstances, the Region proposes to go ahead with project appraisal prior to the availability of the Project Details, the CD director should bring the matter to the attention of the OPNSV (using the procedure described in para. 4 above), giving all the relevant facts related to the international aspects of the project and seeking approval to proceed with appraisal.

7. The other riparians should be allowed a reasonable period of time, which should not normally exceed six months from the dispatch of the Project Details, to communicate their response to the beneficiary state or the Bank.

8. Notification to riparians will not be required in the following cases:

(a) Projects involving additions or alterations by way of rehabilitation, construction, or otherwise to any ongoing schemes that in the judgment of the Bank meet the following criteria:

(i) they will not adversely change the quality or quantity of water flows to other riparians; and

(ii) they will not be adversely affected by the use of water that other riparians might make.

However, if there is any agreement or arrangement between the riparians, Bank staff should make efforts to secure compliance with the requirements of the agreement or arrangement.

(b) Water resource surveys and feasibility studies on or involving international waterways. Beneficiary states should, however, be required to include in the terms of reference for such surveys and studies, an examination of any potential riparian issues.

9. The Bank should ascertain whether the riparians have established any institutional framework with respect to the waterways concerned, and if so, the scope of its activities, functions, and the status of its involvement in the proposed project, bearing in mind the possible need for notifying the institution.

Responses/Objections

10. If the beneficiary state or the Bank, after giving notice, receives from the other riparians a positive response (which may be in the form of, for example,

consent, no objection, support to the project, or a confirmation that the project will not be harmful its interests), or if the other riparians have not responded within the stipulated time limit, the CD director should, in consultation with the Legal Department and the other departments concerned, address a memorandum through the RVP to the OPNSV, giving all the relevant facts (including Bank staff assessment of whether the project will cause appreciable harm to the interests of other riparians, or be so harmed by the use of water that other riparians might make), and seeking approval for further actions.

11. If the other riparians have raised objections to the proposed project, the CD director, in close collaboration with the Legal Department and in consultation with other departments concerned, should address a memorandum through the RVP to the OPNSV and copied to the Vice President and General Counsel, detailing the following:

(a) the nature of the riparian issues;

(b) Bank staff assessment of the objections raised, including the reasons for them and supporting data, as appropriate;

(c) Bank staff assessment of whether the proposed project will cause appreciable harm to the interests of other riparians, or be so harmed by the use of water that other riparians might make;

(d) whether the circumstances of the case require that the Bank should, before taking any further action, urge the parties to resolve the issues through amicable means such as consultations, negotiations, good offices, etc. (which will normally be resorted to, when objections by the other riparians are substantiated); and

(e) whether the objections are of such a nature that it would be advisable to obtain an additional opinion from independent experts in accordance with the provisions of Annex A.

12. The OPNSV, in consultation with the Vice President and General Counsel, will decide on whether and/or how to proceed. In cases where he considers it appropriate, he may request the Operations Committee to consider the matter before he issues his instructions. The outcome of actions taken as a result of these instructions will be reported by the CD director in a memorandum through the RVP to the OPNSV and copied to the Vice President and General Counsel. The memorandum should be prepared in accordance with the procedures in para. 11, and should include appropriate recommendations for further project processing.

13. If independent experts are asked to give their opinion, their conclusions will be reviewed by the OPNSV before a decision is made, in consultation with the Vice President and General Counsel, on whether to proceed with further processing of the project.

14. Should the Bank decide to proceed with the project, despite the objections of other riparians, the Bank will inform those riparians of its decision.

Presentation of Loans to the Executive Directors

15. For every project on international waterways, the Staff Appraisal Report and the Memorandum and Recommendation of the President (MOP) should deal with the international aspects of the project, and should state that Bank staff have considered these aspects and are satisfied that:

(a) the issues involved are covered by appropriate agreement or arrangement between the beneficiary state and other riparians; or

(b) the other riparians have given a positive response to the beneficiary state or to the Bank, in the form of consent, no objection, support to the project, or confirmation that the project will not be harmful to their interests; or

(c) in all other cases, the assessment of Bank staff, the project would not cause appreciable harm to the other riparians, or would not be so harmed by the use of waters by other riparians. The MOP should also contain in an annex the salient features of any objection and, where applicable, the report and conclusions of the independent experts.

Annex A

TECHNICAL ADVICE OF INDEPENDENT EXPERTS

1. Whenever the Senior Vice President, Operations (OPNSV), decides to seek an additional opinion, in accordance with the provisions of this directive, the OPNSV will request the Vice President, Sector Policy and Research (PRSVP), to seek the advice of one or more independent experts for this purpose.

2. The PRSVP will maintain, in consultation with the Regional vice presidents (RVPs) and the Legal Department, a roster of ten independent and highly qualified experts, which will be updated annually.

3. When so requested, the PRSVP will, in consultation with the RVPs concerned and the Legal Department, select the experts from the roster. The experts should not be nationals of any of the riparians of the waterways involved, and should have no conflict of interest in the matter. The experts should be engaged and terms of reference prepared jointly by the offices of the PRSVP and the RVP of the Region processing the project. The latter is responsible for financing the cost associated with the engagement of experts.

4. The experts will be provided with all necessary background information and any assistance required for the efficient completion of their work.

5. The experts shall examine the Project Details. If they deem it necessary to verify the Project Details and take any related action, the Bank will use its best efforts to facilitate this. The experts will meet on an *ad hoc* basis until they submit their report. Upon completion of their work, they will furnish their report and conclusions to the PRSVP and the RVP of the Region concerned, for submission to the OPNSV. The PRSVP or the RVP concerned may ask the experts to provide an explanation or clarification of any point or aspect of their report.

6. The experts shall not have any decision-making role with respect to the processing of the project. Their technical opinion will be submitted only for the Bank's purpose, and shall not in any way determine the rights and obligations of riparians.

OP 7.50
June 2001

OPERATIONAL POLICIES

PROJECTS ON INTERNATIONAL WATERWAYS

These policies were prepared for use by World Bank staff and are not necessarily a complete treatment of the subject.

This Operational Policy statement was revised in August 2004 to reflect the term "development policy lending" (formerly adjustment lending), in accordance with OP/BP 8.60, issued in August 2004.

Note: OP and BP 7.50 replace OP and BP 7.50, dated October 1994. Questions may be addressed to the Chief Counsel, Environmental and International Law.

Applicability of Policy

1. This policy applies to the following types of international waterways:

    (a) any river, canal, lake, or similar body of water that forms a boundary between, or any river or body of surface water that flows through, two or more states, whether Bank[1] members or not;

    (b) any tributary or other body of surface water that is a component of any waterway described in (a) above; and

    (c) any bay, gulf, strait, or channel bounded by two or more states or, if within one state, recognized as a necessary channel of communication between the open sea and other states—and any river flowing into such waters.

2. This policy applies to the following types of projects:

    (a) hydroelectric, irrigation, flood control, navigation, drainage, water and sewerage, industrial, and similar projects that involve the use or potential pollution of international waterways as described in para. 1 above; and

---

[1] "Bank" includes IBRD and IDA; "loans" include IDA credits and IDA grants; and "project" includes all projects financed under Bank loans or IDA credits, but does not include development policy lending programs supported under Bank loans and IDA credits; and "borrower" refers to the member country in whose territory the project is carried out, whether or not the country is the borrower or the guarantor.

(b) detailed design and engineering studies of projects under para. 2(a) above, including those to be carried out by the Bank as executing agency or in any other capacity.

Agreements/Arrangements

3. Projects on international waterways may affect relations between the Bank and its borrowers and between states (whether members of the Bank or not). The Bank recognizes that the cooperation and goodwill of riparians is essential for the efficient use and protection of the waterway. Therefore, it attaches great importance to riparians' making appropriate agreements or arrangements for these purposes for the entire waterway or any part thereof. The Bank stands ready to assist riparians in achieving this end. In cases where differences remain unresolved between the state proposing the project (beneficiary state) and the other riparians, prior to financing the project the Bank normally urges the beneficiary state to offer to negotiate in good faith with the other riparians to reach appropriate agreements or arrangements.

Notification

4. The Bank ensures that the international aspects of a project on an international waterway are dealt with at the earliest possible opportunity. If such a project is proposed, the Bank requires the beneficiary state, if it has not already done so, formally to notify the other riparians of the proposed project and its Project Details (see BP 7.50, para. 3). If the prospective borrower indicates to the Bank that it does not wish to give notification, normally the Bank itself does so. If the borrower also objects to the Bank's doing so, the Bank discontinues processing of the project. The executive directors concerned are informed of these developments and any further steps taken.

5. The Bank ascertains whether the riparians have entered into agreements or arrangements or have established any institutional framework for the international waterway concerned. In the latter case, the Bank ascertains the scope of the institution's activities and functions and the status of its involvement in the proposed project, bearing in mind the possible need for notifying the institution.

6. Following notification, if the other riparians raise objections to the proposed project, the Bank in appropriate cases may appoint one or more independent experts to examine the issues in accordance with BP 7.50, paras. 8–12. Should the Bank decide to proceed with the project despite the objections of the other riparians, the Bank informs them of its decision.

Exceptions to Notification Requirement

7. The following exceptions are allowed to the Bank's requirement that the other riparian states be notified of the proposed project:

(a) For any ongoing schemes, projects involving additions or alterations that require rehabilitation, construction, or other changes that in the judgment of the Bank

(i) will not adversely change the quality or quantity of water flows to the other riparians; and

(ii) will not be adversely affected by the other riparians' possible water use.

This exception applies only to minor additions or alterations to the ongoing scheme; it does not cover works and activities that would exceed the original scheme, change its nature, or so alter or expand its scope and extent as to make it appear a new or different scheme. In case of doubt regarding the extent to which a project meets the criteria of this exception, the executive directors representing the riparians concerned are informed and given at least two months to reply. Even if projects meet the criteria of this exception, the Bank tries to secure compliance with the requirements of any agreement or arrangement between the riparians.

(b) Water resource surveys and feasibility studies on or involving international waterways. However, the state proposing such activities includes in the terms of reference for the activities an examination of any potential riparian issues.

(c) Any project that relates to a tributary of an international waterway where the tributary runs exclusively in one state and the state is the lowest downstream riparian, unless there is concern that the project could cause appreciable harm to other states.

Presentation of Loans to the Executive Directors

8. The Project Appraisal Document (PAD) for a project on an international waterway deals with the international aspects of the project, and states that Bank staff have considered these aspects and are satisfied that:

(a) the issues involved are covered by an appropriate agreement or arrangement between the beneficiary state and the other riparians; or

(b) the other riparians have given a positive response to the beneficiary state or Bank, in the form of consent, no objection, support to the project, or confirmation that the project will not harm their interests; or

(c) in all other cases, in the assessment of Bank staff, the project will not cause appreciable harm to the other riparians, and will not be appreciably harmed by the other riparians' possible water use. The PAD also contains in an annex the salient features of any objection and, where applicable, the report and conclusions of the independent experts.

APPENDIX 5B

BP 7.50
June 2001

BANK PROCEDURES

PROJECTS ON INTERNATIONAL WATERWAYS

These policies were prepared for use by World Bank staff and are not necessarily a complete treatment of the subject.

This Bank Procedures statement was revised in August 2004 to reflect the term "development policy lending" (formerly adjustment lending), in accordance with OP/BP 8.60, issued in August 2004.

Note: OP and BP 7.50 replace OP and BP 7.50, dated October 1994. Questions may be addressed to the Chief Counsel, Environmental and International Law.

1. A potential international water rights issue is assessed as early as possible during project identification[1] and described in all project documents starting with the Project Information Document (PID). The task team (TT) prepares the project concept package, including the PID, in collaboration with the Legal Vice Presidency (LEG) to convey all relevant information on international aspects of the project. When the TT sends the project concept package to the Regional Vice President (RVP), it sends a copy to the Vice President and General Counsel (LEGVP). Throughout the project cycle the Region, in consultation with LEG, keeps the managing director (MD) concerned abreast of the international aspects of the project and related events.

Notification

2. As early as possible during identification, the Bank[2] advises the state proposing the project on an international waterway (beneficiary state) that, if it has not

---

[1] See BP 10.00, *Investment Lending: Identification to Board Presentation*.

[2] "Bank" includes IBRD and IDA; "loans" include IDA credits and IDA grants; and "projects" includes all projects financed under Bank loans or IDA credits, but does not include development policy lending programs supported under Bank loans and IDA credits; and "borrower" refers to the member country in whose territory the project is carried out, whether or not the country is the borrower or the guarantor.

already done so, it should formally notify the other riparians of the proposed project giving available details (see para. 3). If the prospective borrower indicates to the Bank that it does not wish to give notification, normally the Bank itself does so. If the beneficiary state also objects to the Bank's doing so, the Bank discontinues processing of the project. The Region informs the executive directors concerned of these developments and of any further steps taken.

3. The notification contains, to the extent available, sufficient technical specifications, information, and other data (Project Details) to enable the other riparians to determine as accurately as possible whether the proposed project has potential for causing appreciable harm through water deprivation or pollution or otherwise. Bank staff should be satisfied that the Project Details are adequate for making such a determination. If adequate Project Details are not available at the time of notification, they are made available to the other riparians as soon as possible after the notification. If, in exceptional circumstances, the Region proposes to go ahead with project appraisal before Project Details are available, the country director (CD), via a memorandum prepared in consultation with LEG and copied to the LEGVP, notifies the RVP of all relevant facts on international aspects and seeks approval to proceed. In making this decision, the RVP seeks the advice of the MD concerned.

4. The other riparians are allowed a reasonable period, normally not exceeding six months from the dispatch of the Project Details, to respond to the beneficiary state or Bank.

Responses/Objections

5. After giving notice, if the beneficiary state or Bank receives a positive response from the other riparians (in the form of consent, no objection, support to the project, or confirmation that the project will not harm their interests), or if the other riparians have not responded within the stipulated time, the CD, in consultation with LEG and other departments concerned, addresses a memorandum to the RVP. The memorandum reports all relevant facts, including staff assessment of whether the project would (a) cause appreciable harm to the interests of the other riparians, or (b) be appreciably harmed by the other riparians' possible water use. The memorandum seeks approval for further action. In making this decision, the RVP seeks the advice of the MD concerned.

6. If the other riparians object to the proposed project, the CD, in collaboration with LEG and other departments concerned, sends a memorandum on the objections to the RVP and copies it to the LEGVP. The memorandum addresses

(a) the nature of the riparian issues;

(b) the Bank staff's assessment of the objections raised, including the reasons for them and any available supporting data;

(c) the staff's assessment of whether the proposed project will cause appreciable harm to the interests of the other riparians, or be appreciably harmed by the other riparians' possible water use;

(d) the question of whether the circumstances of the case require that the Bank, before taking any further action, urge the parties to resolve the issues through amicable means such as consultations, negotiations, and good offices (which will normally be resorted to when the other riparians' objections are substantiated); and

(e) the question of whether the objections are of such a nature that it is advisable to obtain an additional opinion from independent experts in accordance with paras. 8–12.

7. The RVP seeks the advice of the MD concerned and the LEGVP, and decides whether and how to proceed. On the basis of these consultations, the RVP may recommend to the MD concerned that the Operations Committee consider the matter. The CD then acts upon either the Operations Committee's instructions, which are issued by the chairman, or the RVP's instructions, and reports the outcome in a memorandum prepared in collaboration with LEG and other departments concerned. The memorandum, sent to the RVP and copied to the LEGVP, includes recommendations for processing the project further.

Seeking the Opinion of Independent Experts

8. If independent expert opinion is needed before further processing of the project (see OP 7.50, para. 6), the RVP requests the Vice President, Sustainable Development Network (SDNVP) to initiate the process. The Office of the SDNVP maintains a record of such requests.

9. The SDNVP, in consultation with the RVP and LEG, selects one or more independent experts from a roster maintained by SDNVP (see para. 12). The experts selected may not be nationals of any of the riparians of the waterways in question, and also may not have any other conflicts of interest in the matter. The experts are engaged and their terms of reference prepared jointly by the offices of the SDNVP and the RVP. The latter finances the costs associated with engaging the experts. The experts are provided with the background information and assistance needed to complete their work efficiently.

10. The experts' terms of reference require that they examine the Project Details. If they deem it necessary to verify the Project Details or take any related action, the Bank makes its best efforts to assist. The experts meet on an ad hoc basis until they submit their report to the SDNVP and the RVP. The SDNVP or RVP may ask them to explain or clarify any aspect of their report.

11. The experts have no decision-making role in the project's processing. Their technical opinion is submitted for the Bank's purposes only, and does not in any

way determine the rights and obligations of the riparians. Their conclusions are reviewed by the RVP and SDNVP, in consultation with the LEGVP.

12. The SDNVP maintains, in consultation with the RVPs and LEG, the roster of highly qualified independent experts, which consists of 10 names and is updated at the beginning of each fiscal year.

Maps

13. Documentation for a project on an international waterway includes a map that clearly indicates the waterway and the location of the project's components. This requirement applies to the PAD, the Project Information Document (PID), and any internal memoranda that deal with the riparian issues associated with the project. Maps are provided for projects on international waterways even when notification to riparians is not required by the provisions of OP 7.50. Maps are prepared and cleared in accordance with Administrative Manual Statement 9.50, Cartographic Services, and its annexes.

14. However, the inclusion of maps in the cited documents, except internal memoranda, is subject to any general instruction or decision of the Regional Vice President, taken in consultation with the Vice President and General Counsel, to omit maps of the beneficiary state in their entirety or in part.

GOOD PRACTICES

PROJECTS ON INTERNATIONAL WATERWAYS

Good Practices statements (GPs) are advisory. This GP contains information that World Bank staff may find useful in carrying out the Bank's policies and procedures. It is not necessarily a complete treatment of the subject.

1. When changes occur in international boundaries, some surface waters that formerly were national in character become international waterways, requiring increased vigilance in identifying riparian issues. Regional staff assigned to handle any project covered in OP 7.50, para. 2—whether financed by the Bank, the Global Environment Facility, or any trust fund—should immediately check whether the surface waters involved are of international character. When in doubt, Regional staff should check with the lawyer concerned in the Legal Vice Presidency (LEG).

2. Often, state authorities advised by Bank staff to notify other riparians in accordance with OP 7.50 have questioned aspects of and sought the reasons for the Bank's riparian policy. In responding, staff should seek the assistance of the Chief Counsel, Environmental and International Law, who should join in discussions with the state authorities.

3. Every effort should be made to allow the notified riparians six months to respond to the notification. A lesser period is advisable only in cases of emergency.

4. Upon being notified that a riparian may seek additional information or clarification, staff should make every effort to provide it and allow a reasonable period for study and response.

5. When the Office of the Vice President, Sustainable Development Network (SDNVP) updates the roster of independent experts each fiscal year (see BP 7.50, paras. 8–12), the roster is communicated to the Senior Vice President and General Counsel (LEGVP).

Bibliography

Alam, Undala Z., *Water Rationality: Mediating the Indus Waters Treaty*, unpublished Ph.D. Thesis (University of Durham, United Kingdom 1998).

Andersen, Inger *et al, The Niger River Basin—A Vision for Sustainable Management* (The World Bank 2005).

Barberis, Julio A., *International Groundwater Resources Law*, FAO Legislative Study No. 40 (FAO 1986).

Baxter, R. R., *The Law of International Waterways* (Harvard University Press 1964);

Berber, F. J., *Rivers in International Law* (Stevens & Sons 1959).

Bevans, Charles, (ed.), *Treaties and Other International Agreements of the United States of America 1776-1949* (1968–1976).

Blaustein, Albert P., & Gisbert H. Flanz (eds.), *Constitutions of the Countries of the World* (Oceana Publications 2006).

Boisson de Chazournes, Laurence & Salman M. A. Salman, (ed.), *Water Resources and International Law* (The Hague Academy of International Law 2005).

Bogdanovic, Slavko, *International Law of Water Resources—Contribution of the International Law Association* (Kluwer Law International 2001).

Bourne, Charles, *International Water Law—Selected Writings of Professor Charles Bourne* (Patricia Wouters ed., Kluwer Law International 1997).

———, *The International Law Association's Contribution to International Water Resources Law*, 36 Nat. Resources J. 155 (1996).

Browder, Greg & Leonard Ortolano, *The Evolution of an International Water Resources Management Regime in the Mekong River Basin*, 40 Natural Resources Journal 499 (2000).

Brownlie, Ian, *African Boundaries—A Legal and Diplomatic Encyclopedia* (C. Hurst & Company 1979).

Burhenne, W. E., (ed.), *International Environmental Law: Multilateral Treaties* (Kluwer Law International 1996).

Caflisch, Lucius, *Regulation of the Uses of International Watercourses*, in *International Watercourses—Enhancing Cooperation and Managing Conflict*,

World Bank Technical Paper No. 414 (Salman M. A. Salman & Laurence Boisson de Chazournes, eds., The World Bank 1998).

Caponera, Dante, *National and International Water Law and Administration— Selected Writings* (Kluwer Law International 2003).

Chauhan, B. R., *Settlement of International and Inter-state Water Disputes in India* (India Law Institute 1992).

Collins, Robert O., *The Nile* (Yale University Press 2002).

Crow, Ben, *et al., Sharing the Ganges: The Politics and Technology of River Development* (Sage Publications, 1995).

del Castillo-Laborde, Lilian, *The Río de la Plata and its Maritime Front Legal Regime* (Martinus Nijhoff Publishers 2008).

Di Leva, Charles E., *Access to Information, Public Participation and Conflict Resolution at the World Bank*, in *Public Participation in the Governance of International Freshwater Resources* (Carl Bruch *et al.*, eds., United Nations University Press 2005).

Downing, David, *An Atlas of Territorial and Border Disputes* (New English Library 1980).

Dubicki, Alfred & Andrzej Nalberczyński, *Development of the International Cooperation in the Oder River Basin* in *Management of Transboundary Waters in Europe* (Malgorzata Landsberg-Uczciwek, Martin Adriaanse & Rainer Enderlein, eds. 1998).

Eckstein, Gabriel E., *Commentary on the U.N. International Law Commission's Draft Articles on the Law of Transboundary Aquifers*, 18 Colo. J. Int'l Envt. L & Pol'y, 537 (2007).

———, *A Hydrogeological Perspective of the Status of Ground Water Resources Under the UN Watercourse Convention*, 30 Columbia Journal of Environmental Law 525 (2005).

Economic Commission for Europe, *Our Waters: Joining Hands Across Borders— First Assessment of Transboundary Rivers, Lakes and Groundwaters* (United Nation 2007).

Farrajota, Maria Manuela, *Notification and Consultation in the Law Applicable to International Watercourses*, in *Water Resources and International Law* (L. Boisson de Chazournes & S. Salman, eds., The Hague Academy of International Law 2005).

Food and Agriculture Organization, *Groundwater in International Law— Compilation of Treaties and Other Legal Instruments*, FAO Legislative Study 86 (FAO 2005).

———, *Treaties Concerning the Non-Navigational Uses of International Watercourses—Africa*, 61 FAO Legislative Series (FAO 1997).

Freestone, David, & Salman M. A. Salman, *Ocean and Freshwater Resources*, in *The Oxford Handbook of International Environmental Law* (Daniel Bodansky, Jutta Brunnee & Ellen Hey, eds., Oxford University Press 2007).

Freestone, David, *The Environmental and Social Safeguard Policies of the World Bank and the Evolving Role of the Inspection Panel*, in *Economic Globalization and Compliance with International Environmental Agreements* (Kanami Ishibashi, Alexandre Kiss & Dinah Shelton, eds., Kluwer Law International 2003).

Freestone, David, & Ellen Hey (eds.), *The Precautionary Principle and International Law: The Challenge of Implementation* (Kluwer Law International 1996).

Garretson, A. H., *et al.*, (eds.), *The Law of International Drainage Basins* (Oceana Publications 1967).

Getches, David, *Water Law in a Nutshell* (West Publishing 1997).

Godana, Bonaya Adhi, *Africa's Shared Water Resources—Legal and Institutional Aspects of the Nile, Niger and Senegal River Systems* (Francis Pinter Publishers 1985).

Goldberg, David, *Legal Aspects of the World Bank Policy on Projects on International Waterways*, 7 International Journal of Water Resources Development 225 (1991).

Grewe, Wilhelm G., (ed.), *Fontes Historiale Iuris Gentium, (Sources Relating to the History of the Law of Nations)* (1992).

Gulhati, Niranjan D., *Indus Waters Treaty—An Exercise in International Mediation* (Allied Publishers 1973).

Hayton, Robert D., & Albert E. Utton, *Transboundary Groundwaters: The Bellagio Draft Treaty*, 29 Nat. Resources J. 663 (1989).

Heyns, Piet, *Water Resources Management in Southern Africa*, in *International Waters in Southern Africa* (Mikiyasu Nakayama, ed., United Nations University Press 2003).

Hillel, Daniel, *Rivers of Eden—The Struggle for Water and the Quest for Peace in the Middle East* (Oxford University Press 1994).

Hohmann, Harald, *Basic Documents of International Environmental Law*, vol. 1 (Springer 1992).

Howell, P. P. & J. A. Allan (eds.), *The Nile—Resource Evaluation, Resource Management, Hydropolitics and Legal Issues* (School of African and Oriental Studies, University of London 1990).

Hunter, David, James Salzman & Durwood Zaelke, *International Environmental Law and Policy* (Foundation Press 1998).

Hurst, Sir Cecil, *The Territoriality of Bays*, 3 Brit. Yb. Int'l L. 42 (1922–23).

IIL, *Annuaire de l'Institut de Droit International*, 1911–1997.

ILA, *Principles of Law and Recommendations on the Uses of International Rivers* (London 1959).

———, *Conference Reports*, 1956–2004.

ILC, Reports, 1976–2008.

International Bank for Reconstruction and Development and International Development Association, *Annual Reports*, 1965–1966.

International Bank for Reconstruction and Development, *Annual Reports*, 1946–1960.

International Consortium for Cooperation on the Nile (ICCON), *Nile Basin Initiative: Strategic Action Program—Overview* (Prepared by the Nile Basin Initiative Secretariat in Cooperation with the World Bank, May 2001).

Kibaroglu, Aysegül, *Building a Regime for the Waters of the Euphrates-Tigris River Basin* (Kluwer Law International 2002).

Kirmani, Syed, & Guy Le Moigne, *Fostering Riparian Cooperation in International River Basins: The World Bank at its Best in Development Diplomacy*, World Bank Technical Paper No. 335 (The World Bank 1997).

Kolars, John, *Problems of International River Management: The Case of the Euphrates*, in *International Waters of the Middle East—From Euphrates to Nile* (Asit Biswas, ed., Oxford University Press 1994).

Krishna, Raj, & Salman M. A. Salman, *International Groundwater Law and the World Bank Policy for Projects on Transboundary Groundwater*, in *Groundwater: Legal and Policy Perspectives*, World Bank Technical Paper No. 456 (Salman M. A. Salman, ed., The World Bank 1999).

Krishna, Raj, *The Evolution and Context of the Bank Policy for Projects on International Waterways*, in *International Watercourses—Enhancing Cooperation and Managing Conflict*, World Bank Technical Paper No. 414 (Salman M. A. Salman & Laurence Boisson de Chazournes, eds., The World Bank 1998).

———, *International Watercourses: World Bank Experience and Policy*, in *Water in the Middle East: Legal, Political and Commercial Implications* (J. A. Allan & Chibli Mallat, eds., I. B. Tauris Publishers 1995).

———, *International Waters in Bank Projects—History and Case Studies*, Legal Department Working Paper No. 2 (The World Bank 1973).

Lipper, Jerome, *Equitable Utilization* in *The Law of International Drainage Basins* (Albert Garretson, *et al.* eds., Oceana 1967).

Lowi, Miriam, *Water and Power—The Politics of a Scarce Resource in the Jordan River Basin* (Cambridge University Press, 1995).

Macoun, Andrew & Hazim El Naser, *Groundwater Resources Management in Jordan: Policy and Regulatory Issues*, in *Groundwater: Legal and Policy Perspectives*, World Bank Technical Paper No. 456 (Salman M. A. Salman, ed., The World Bank 1999).

Mason, Edward S. & Robert E. Asher, *The World Bank since Bretton Woods* (The Brookings Institution 1973).

McCaffrey, Stephen, *The Law of International Watercourses* (2d. ed., Oxford University Press 2007).

————, *Water Disputes Defined: Characteristics and Trends for Resolving Them*, in *Resolution of International Water Disputes* (International Bureau of the Permanent Court of Arbitration, Kluwer Law International 2003).

————, *International Groundwater Law: Evolution and Context*, in *Groundwater: Legal and Policy Perspectives*, World Bank Technical Paper No. 456 (Salman M. A. Salman, ed., The World Bank 1999).

————, *The UN Convention on the Law of the Non-Navigational Uses of International Watercourses: Prospects and Pitfalls*, in *International Watercourses—Enhancing Cooperation and Managing Conflict*, World Bank Technical Paper No. 414 (Salman M. A. Salman & Laurence Boisson de Chazournes, eds., The World Bank 1998).

————, *The Harmon Doctrine One Hundred Years Later: Buried, Not Praised*, 36 Nat. Resources J. 549 (1996).

Meital, Yoram, *The Aswan High Dam and Revolutionary Symbolism in Egypt*, in *The Nile—Histories, Cultures, Myths* (Haggai Erlich & Israel Gershoni, eds., Lynne Rienner Publishers 2000).

Michel, Aloys Arthur, *The Indus River—A Study of the Effects of Partition* (Yale University Press 1967).

Okidi , C. O., *History of the Nile and Lake Victoria Basins Through Treaties*, in *The Nile—Resources Evaluation, Resource Management, Hydropolitics and Legal Issues* (P. P. Howell & J. A. Allan, eds., School of African and Oriental Studies, University of London 1990).

Paisley, Richard, *Adversaries into Partners: International Water Law and the Equitable Sharing of Downstream Benefits*, 13 Melbourne Journal of International Law 280 (2002).

Redgwell, Catherine, *From Permission to Prohibition: The 1982 Convention on the Law of the Sea and Protection of the Marine Environment*, in David Freestone, Richard Barnes & David M. Ong, *The Law of the Sea—Progress and Prospects* (Oxford University Press 2006).

Salman, Salman M. A., & Daniel Bradlow, *Regulatory Frameworks for Water Resources Management—A Comparative Study* (The World Bank 2006).

Salman, Salman M. A., & Kishor Uprety, *Hydro-Politics in South Asia: A Comparative Analysis of the Mahakali and the Ganges Treaties*, 39 Natural Resources Journal (1999).

————, *Conflict and Cooperation on South Asia's International Rivers—A Legal Perspective* (Kluwer Law International 2002).

Salman, Salman M. A., *The Baglihar Difference and its Resolution Process—A Triumph for the Indus Waters Treaty?* 10 Water Policy 105 (2008).

———, *The Niger River*, in *Max Plank Encyclopedia of Public International Law* (Oxford University Press 2009).

———, *The Helsinki Rules, the United Nations Watercourses Convention and the Berlin Rules*, 23 International Journal of Water Resources Development 525 (2007).

———, *The United Nations Watercourses Convention Ten Years Later—Why Has its Entry into Force Proven Difficult?* 32 Water International 1 (2007).

———, *Good Offices, Mediation and International Water Disputes*, in Resolution of International Water Disputes (The International Bureau of the Permanent Court of Arbitration, Kluwer Law International 2003).

———, *Inter-states Water Disputes in India—An Analysis of the Settlement Process*, 4 Water Policy (2002).

———, *Dams, International Rivers, and Riparian States: An Analysis of the Recommendations of the World Commission on Dams*, 16(6) Am. U. Int'l L. Rev. 1477 (2001).

———, *Legal Regime for Use and Protection of International Watercourses in the Southern African Region: Evolution and Context*, 41 Natural Resources Journal (2001).

———, *International Rivers as Boundaries: The Dispute Over The Kasikili/Sedudu Island and the Decision of the International Court of Justice*, 25 Water International 580 (2000).

———, *Groundwater: Legal and Policy Perspectives*, World Bank Technical Paper No. 456 (The World Bank 1999).

Shapland, Greg, *Rivers of Discord—International Water Disputes in the Middle East* (Hurst & Company 1997).

Shihata, Ibrahim, *The World Bank Inspection Panel* (2d ed., Oxford 2001).

———, *Foreword* to *International Watercourses—Enhancing Cooperation and Managing Conflict*, World Bank Technical Paper No. 414 (Salman M. A. Salman & Laurence Boisson de Chazournes, eds., The World Bank 1998).

———, *The World Bank and the Environment: A Legal Perspective*, 16 MD. J. Int'l L & TR 1, (4) 1992).

———, *The World Bank in a Changing World: Selected Essays and Lectures* (Martinus Nijhoff 1991).

Smith, Herbert Arthur, *The Economic Uses of International Rivers* (P. S. King & Son, Ltd. 1931).

Soffer, Arnon, *Rivers of Fire: The Conflict of Water in the Middle East* (Rowman & Littlefield Publishers 1999).

Stein, Robyn, *South Africa's New Democratic Water Legislation: National Government's Role as a Public Trustee in Dam Building and Management Activities*, 18 Journal of Energy and Natural Resources Law 284 (2000).

Subedi, Surya, *Resolution of International Water Disputes: Challenges for the 21st Century*, in *Resolution of International Water Disputes* (The International Bureau of the Permanent Court of Arbitration, ed., Kluwer Law International 2003).

Sutcliffe, J. V., & Y. P Parks, *The Hydrology of the Nile*, IAHS Special Publication No. 5 (International Association of Hydrological Sciences 1999).

Tanzi, Attila, & Maurizio Arcari, *The United Nations Convention on the Law of International Watercourses* (Kluwer Law International 2001).

Teclaff, Ludwik A., *Water Law in Historical Perspective* (William S. Hein & Co. 1985).

Tsutsumi, Rie, & Kristy Robinson, *Environmental Impact Assessment and the Framework Convention for the Protection of the Marine Environment of the Caspian Sea*, in *Theory and Practice of Transboundary Environmental Impact Assessment* (Kees Bastmeijer & Timo Koivurova, eds., Martinus Nijhoff Publishers 2008).

van Putten, Maartje, *Policing the Banks—Accountability Mechanisms for the Financial Sector* (McGill-Queen's University Press 2008).

Vitanyi, Béla, *The International Regime of River Navigation* (Kluwer Law International 1979).

von Martens, Karl, & Ferdinand de Cussey, *Recueil Manuel et Pratique de Traités, Conventions et Autres Actes Diplomatiques* (1935).

Waterbury, John, *The Nile Basin—National Determinants and Collective Action* (Yale University Press 2000).

Watts, Sir Arthur, *The International Law Commission* 1949–98, Volume Two: The Treaties (Oxford University Press 1999).

World Bank, *Making the Most of Scarcity, Accountability for Better Water Management in the Middle East and North Africa* (The World Bank 2007).

―――, *Water Resources Sector Strategy—Strategic Directions for World Bank Engagement* (The World Bank, 2004).

―――, *Environmental Assessment Sourcebook* (The World Bank 1999).

―――, *Pollution Prevention and Abatement Handbook* (The World Bank 1999).

―――, *From Scarcity to Security—Averting a Water Crisis in the Middle East and North Africa* (The World Bank 1998).

―――, *Water Resources Management—A World Bank Policy Paper* (The World Bank 1993).

―――, *Annual Reports*, 1985–2008.

World Commission on Dams, *Dams and Development: A New Framework for Decision-Making* (Earthscan Publications, London 2000).

Index

www.ingramcontent.com/pod-product-compliance
Lightning Source LLC
Chambersburg PA
CBHW070717280326
41926CB00087B/2397